3

COLLOQUIA

Michael G. Cowling
School of Mathematics and Statistics
University of New South Wales
Sydney NSW 2052, Australia

Joseph A. Wolf
Department of Mathematics
University of California
Berkeley
California 94720–3840, U.S.A.

Gisbert Wüstholz
Departement Mathematik
ETH Zürich
Rämistrasse 101
CH-8092 Zürich, Switzerland

David Mumford
Division of Applied Mathematics
Brown University
182 George Street
Providence
RI 02912, USA

Colloquium De Giorgi 2009

Colloquium
De Giorgi
2009

edited by
Umberto Zannier

EDIZIONI
DELLA
NORMALE

© 2012 Scuola Normale Superiore Pisa

ISBN 978-88-7642-388-8
ISBN 978-88-7642-387-1 (eBook)

Contents

Preface

Since 2001 the Scuola Normale Superiore di Pisa has organized the "Colloquio De Giorgi", a series of colloquium talks named after Ennio De Giorgi, the eminent analyst who was a Professor at the Scuola from 1959 until his death in 1996.

The Colloquio takes place once a month. It is addressed to a general mathematical audience, and especially meant to attract graduate students and advanced undergraduate students. The lectures are intended to be not too technical, in fields of wide interest. They should provide an overview of the general topic, possibly in a historical perspective, together with a description of more recent progress.

The idea of collecting the materials from these lectures and publishing them in annual volumes came out recently, as a recognition of their intrinsic mathematical interest, and also with the aim of preserving memory of these events.

For this purpose, the invited speakers are now asked to contribute with a written exposition of their talk, in the form of a short survey or extended abstract. This series has been continued in a collection that we hope shall be increased in the future.

This volume contains a complete list of the talks held in the "Colloquio De Giorgi" in 2009 and also in the past years, and a table of contents of the first two volumes too.

Colloquia held in 2001

Paul Gauduchon
Weakly self-dual Kähler surfaces

Tristan Rivière
Topological singularities for maps between manifolds

Frédéric Hélein
Integrable systems in differential geometry and Hamiltonian stationary Lagrangian surfaces

Jean-Pierre Demailly
Numerical characterization of the Kähler cone of a compact Kähler manifold

Elias Stein
Discrete analogues in harmonic analysis

John N. Mather
Differentiability of the stable norm in dimension less than or equal to three and of the minimal average action in dimension less than or equal to two

Guy David
About global Mumford-Shah minimizers

Jacob Palis
A global view of dynamics

Alexander Nagel
Fundamental solutions for the Kohn-Laplacian

Alan Huckleberry
Incidence geometry and function theory

Colloquia held in 2002

Michael Cowling
Generalizzazioni di mappe conformi

Felix Otto
The geometry of dissipative evolution equations

Curtis McMullen
Dynamics on complex surfaces

Nicolai Krylov
Some old and new relations between partial differential equations, stochastic partial differential equations, and fine properties of the Wiener process

Tobias H. Colding
Disks that are double spiral staircases

Cédric Villani
When statistical mechanics meets regularity theory: qualitative properties of the Boltzmann equation with long-range interactions

Colloquia held in 2003

John Toland
Bernoulli free boundary problems - progress and open questions

Jean-Michel Morel
The axiomatic method in perception theory and image analysis

Jacques Faraut
Random matrices and infinite dimensional harmonic analysis

Albert Fathi
C^1 *subsolutions of Hamilton-Iacobi Equation*

Hakan Eliasson
Quasi-periodic Schrödinger operators - spectral theory and dynamics

Yakov Pesin
Is chaotic behavior typical among dynamical systems?

Don B. Zagier
Modular forms and their periods

David Elworthy
Functions of finite energy in finite and infinite dimensions

Colloquia held in 2004

Jean-Christophe Yoccoz
Hyperbolicity for products of 2×2 *matrices*

Giovanni Jona-Lasinio
Probabilità e meccanica statistica

John H. Hubbard
Thurston's theorem on combinatorics of rational functions and its generalization to exponentials

Marcelo Viana
Equilibrium states

Boris Rozovsky
Stochastic Navier - Stokes equations for turbulent flows

Marc Rosso *Braids and shuffles*

Michael Christ
The d-bar Neumann problem, magnetic Schrödinger operators, and the Aharonov-Böhm phenomenon

Colloquia held in 2005

Louis Nirenberg
One thing leads to another

Viviane Baladi
Dynamical zeta functions and anisotropic Sobolev and Hölder spaces

Giorgio Velo
Scattering non lineare

Gerd Faltings
Diophantine equations

Martin Nowak
Evolution of cooperation

Peter Swinnerton-Dyer
Counting rational points: Manin's conjecture

François Golse
The Navier-Stokes limit of the Boltzmann equation

Joseph J. Kohn
Existence and hypoellipticity with loss of derivatives

Dorian Goldfeld
On Gauss' class number problem

Colloquia held in 2006

Yuri Bilu
Diophantine equations with separated variables

Corrado De Concini
Algebre con tracce e rappresentazioni di gruppi quantici

Zeev Rudnick
Eigenvalue statistics and lattice points

Lucien Szpiro
Algebraic Dynamics

Simon Gindikin
Harmonic analysis on complex semisimple groups and symmetric spaces from point of view of complex analysis

David Masser
From 2 to polarizations on abelian varieties

Colloquia held in 2007

Klas Diederich
Real and complex analytic structures

Stanislav Smirnov
Towards conformal invariance of 2D lattice models

Roger Heath-Brown
Zeros of forms in many variables

Vladimir Sverak
PDE aspects of the Navier-Stokes equations

Christopher Hacon
The canonical ring is finitely generated

John Coates
Elliptic curves and Iwasava theory

Colloquia held in 2008

Claudio Procesi
Funzioni di partizione e box-spline

Pascal Auscher
Recent development on boundary value problems via Kato square root estimates

Hendrik W. Lenstra
Standard models for finite fields

Jean-Michel Bony
Generalized Fourier integral operators and evolution equations

Shreeram S. Abhyankar
The Jacobian conjecture

Fedor Bogomolov
Algebraic varieties over small fields

Louis Nirenberg
On the Dirichlet problem for some fully nonlinear second order elliptic equations

Contents of previous volumes:

Isomorphisms of the Figà-Talamanca–Herz Algebras $A_p(G)$ for connected Lie groups G

Michael G. Cowling

Abstract. There are various Banach algebras of functions on a locally compact group G, made up of matrix coefficients of representations, such as the Fourier algebra $A(G)$ and the Fourier–Stieltjes algebra $B(G)$, which reflect the representation theory of the group. The question of whether these determine the group has been considered by many authors. Here we show that when $1 < p < \infty$, the Figà-Talamanca–Herz algebras $A_p(G)$ determine the group G, at least if G is a connected Lie group.

1. Introduction and notation

Denote by G a locally compact Hausdorff topological group, henceforth just called a group, with a left-invariant Haar measure m. We will occasionally write G_d for G with the discrete topology. For group elements, we write x, y, $\ldots$; e denotes the identity. Group maps, including representations, will always be continuous.

Functions on G are complex-valued, unless otherwise stated. The usual Lebesgue spaces are written $L^p(G)$, and $C_0(G)$ is the space of continuous functions on G vanishing at infinity. Functions will be written f, g, $\ldots$, and also u, v, $\ldots$.

1.1. Cosets and affine maps

We will need various generalizations of group isomorphisms; to describe these, it is useful to first discuss cosets.

It is well known that a subset S of a group G is a subgroup if and only if

$$x_1, x_2 \in S \quad \Longrightarrow \quad x_1 x_2^{-1} \in S.$$

Similarly, a subset S of a group G is a coset (left cosets are right cosets of possibly different subgroups, so we do not need to clarify whether a

The author wishes to thank the *Centro di Ricerca Matematica Ennio De Giorgi* and the *Alexander von Humboldt Stiftung* for their support.

coset is a left coset or a right coset) if and only if

$$x_1, x_2, x_3 \in S \quad \Longrightarrow \quad x_1 x_2^{-1} x_3 \in S.$$

The coset ring $\Omega(G)$ of a group G is the ring of subsets of G generated by the cosets of open subgroups of G. By the way, a subset S of a group G is a subgroup if and only if

$$x_1, x_2, \ldots, x_{2N-1}, x_{2N} \in S \quad \Longrightarrow \quad x_1 x_2^{-1} \ldots x_{2N-1}^{-1} x_{2N} \in S$$

and is a coset if and only if

$$x_1, x_2, \ldots, x_{2N}, x_{2N+1} \in S \quad \Longrightarrow \quad x_1 x_2^{-1} \ldots x_{2N}^{-1} x_{2N+1} \in S$$

for any positive integer N. We write CosetMaps(G) for the group of all homeomorphisms ϕ of G for which $\phi(C)$ is a coset of a closed subgroup of G if and only if C is a coset of a closed subgroup of G, and CosetMaps(G, G') for the analogously defined set of maps from G to G'. Clearly, translations are coset-preserving maps, and by composing a coset-preserving map with a translation, we may obtain a new coset-preserving map that sends e to e.

A map of (cosets in) groups $\phi \colon G \to H$ is said to be *affine* if

$$\phi(x_1 x_2^{-1} x_3) = \phi(x_1)\phi(x_2)^{-1}\phi(x_3) \quad \forall x_1, x_2, x_3 \in G. \qquad (1.1)$$

A map of (cosets in) groups $\phi \colon G \to H$ is said to be *extended affine* if (1.1) holds or

$$\phi(x_1 x_2^{-1} x_3) = \phi(x_3)\phi(x_2)^{-1}\phi(x_1) \quad \forall x_1, x_2, x_3 \in G. \qquad (1.2)$$

Obviously, for abelian groups, there is no distinction between affine and extended affine maps, but for nonabelian groups, the inversion map $x \mapsto x^{-1}$ is extended affine but not affine. Group homomorphisms are affine maps, as are translations. It is easy to see that an affine map that sends e to e is a homomorphism, and then deduce that any affine map is composed of a homomorphism and a translation, so the set Aff(G) of all affine homeomorphisms of G is a group, isomorphic to Aut(G) $\ltimes G$. The extended affine group ExtAff(G) of all extended affine homeomorphisms of G is just Aff(G), extended by inversion if G is nonabelian.

There is also a natural notion of *Jordan affine map*, satisfying

$$\phi(x_1 x_2^{-1} x_1) = \phi(x_1)\phi(x_2)^{-1}\phi(x_1) \quad \forall x_1, x_2 \in G.$$

In any group G, the "symmetry" $x \mapsto x^{-1}$ about the identity e extends to a "symmetry" about any point x_1; this symmetry is given by

$x_2 \mapsto x_1 x_2^{-1} x_1$, and Jordan affine maps are precisely the maps that preserve all these "symmetries". Linear Jordan maps of associative algebras are automatically homomorphisms or anti-homomorphisms; Jordan affine maps of some groups are automatically extended affine, but for other groups this is not true. For example, on the one hand, Jordan affine maps of abelian groups in which all elements have square roots are affine, while on the other hand, arbitrary dilations of the $(2n+1)$-dimensional Heisenberg group H^n are Jordan affine, as are the natural maps from $\mathbb{R}^{2n+1}$ to H^n.

0	2	4	6	8
5	7	9	1	3

Figure 1. The cosets in $\mathbb{Z}_{10}$.

Extended affine maps send cosets to cosets. One question that underlines our work is whether a map of groups that sends cosets to cosets is necessarily extended affine. The answer is not always yes, as consideration of the cosets in $\mathbb{Z}_{10}$, the cyclic group of order 10, shows (I thank Rob Curtis who pointed this out to me). These cosets are represented in Figure 1: there are two cosets of order 5 (the horizontal rows) and five cosets of order 2 (the vertical columns). The coset preserving maps sending 0 to 0 correspond with the permutations of $\{2, 4, 6, 8\}$, and there are 24 of these, while the image of 2 under an extended affine map sending 0 to 0 is an element of $\{2, 4, 6, 8\}$, and this image determines the image of 4, of 6 and of 8, and then the images of the odd numbers too; hence there are only 4 such extended affine maps.

1.2. Definition of $A_p(G)$

The left regular representation of G on $L^p(G)$ is written λ_p. For f in $L^p(G)$, we define

$$[\lambda_p(x)f](y) = f(x^{-1}y) \quad \forall x, y \in G.$$

The space $A_p(G)$ is the minimal Banach space of functions generated by matrix coefficients of the regular representation of G on $L^p(G)$. A function u is in $A_p(G)$ if and only if it admits a representation of the form

$$u(x) = \sum_{n \in \mathbb{N}} \langle \lambda_p(x) g_n, h_n \rangle = \sum_{n \in \mathbb{N}} \int_G g_n(x^{-1}y) h(y) \, dy \quad \forall x \in G, \quad (1.3)$$

where $g_n \in L^p(G)$, $h_n \in L^{p'}(G)$, and the following expression is finite:

$$\sum_{n \in \mathbb{N}} \|g_n\|_p \|h_n\|_{p'} . \tag{1.4}$$

We replace $L^\infty(G)$ by $C_0(G)$ when p is either 1 or ∞. The norm of u in $A_p(G)$ is the infimum of all the expressions (1.4) such that (1.3) holds.

Eymard [8] introduced the Fourier algebra $A(G)$ of G, which is just $A_p(G)$ when $p = 2$, as well as the Fourier–Stieltjes algebra $B(G)$, which involves coefficients of all unitary representations of G rather than just the regular representation. When G is abelian, the Fourier transformation identifies $A(G)$ and $L^1(\hat{G})$.

The space $A_p(G)$ was defined (for abelian G) by Alessandro Figà-Talamanca [10], and generalized to general locally compact groups by Carl Herz [12], who also showed that the space $A_p(G)$ is a Banach algebra of functions with pointwise operations. Consequently, $A_p(G)$ is often called the Figà-Talamanca–Herz algebra. For all p in $[1, \infty]$, the space $A_p(G)$ is a dense subspace of $C_0(G)$. If $p = 1$ or $p = \infty$, then $A_p(G) = C_0(G)$. For much more about the algebras $A_p(G)$, see Eymard's survey [9].

For fixed x in G, the evaluation $u \mapsto u(x)$ is a continuous multiplicative linear map from $A_p(G)$ to $\mathbb{C}$. The group G "is" the set of all continuous multiplicative linear functionals on $A_p(G)$. If $\alpha \colon A_p(G') \to A_p(G)$ is a Banach algebra isomorphism (not necessarily isometric) then there is a homeomorphism $\phi \colon G \to G'$ such that $\alpha(u) = u \circ \phi$. However, the Gel'fand theory of commutative Banach algebras does not give the multiplicative structure of G.

We are interested in the question whether the Banach algebra $A_p(G)$ determines G. In general the answer is no; for finite groups G, the algebra $A_p(G)$ coincides with $C(G)$, and for all groups, $A_p(G) = C_0(G)$ isometrically if $p = 1$ or $p = \infty$; in these cases, $A_p(G)$ only determines G as a topological space. However, there are some positive results in this direction. Henceforth we always assume that $1 < p < \infty$.

Theorem 1.1 (P.M. Cohen [4]). *Suppose that G and G' are abelian groups, and $\alpha \colon A(G') \to A(G)$ is an isomorphism. Write $\phi \colon G \to G'$ for the associated homeomorphism. If α is isometric, then ϕ is affine; in general, there is a partition $\bigsqcup_{j=1}^J S_j$ of G into open and closed subsets in $\Omega(G)$, each of which is contained in a coset C_j of a subgroup of G, and affine maps $\phi_j \colon C_j \to G'$, such that ϕ and ϕ_j coincide on the sets S_j for all $j \in \{1, \ldots, J\}$.*

A map satisfying the conclusion of this theorem is said to be "nearly affine".

Theorem 1.2 (M.E. Walter [23]). *Suppose that G and G' are groups, and $\alpha \colon A(G') \to A(G)$ is an isometric isomorphism. Then the associated homeomorphism $\phi \colon G \to G'$ is extended affine.*

Theorem 1.3 (N. Lohoué [18]). *Suppose that G and G' are abelian groups, and $\alpha \colon A_p(G') \to A_p(G)$ is an isometric isomorphism. Then the associated homeomorphism $\phi \colon G \to G'$ is affine.*

Theorem 1.4 (M. Baronti [1]). *Suppose that G and G' are groups, and $\alpha \colon A_p(G') \to A_p(G)$ is an isometric isomorphism. Then the associated homeomorphism $\phi \colon G \to G'$ is Jordan affine.*

Theorem 1.5 (M. Ilie and N. Spronk [16]). *Suppose that G and G' are amenable groups, and $\alpha \colon A(G') \to A(G)$ is a complete isomorphism or a complete isometry. Then the associated homeomorphism $\phi \colon G \to G'$ is nearly affine or affine respectively.*

The proof of Ilie and Spronk, like that of Cohen, uses the fact that a map of $G \times G$ to $G \times G'$ that sends cosets to cosets is affine. The main problem is to produce such a map; this is done by extending the map α to a map from $A(G' \times G)$ to $A(G \times G)$. The "complete" hypothesis in [16] is exactly what is needed to ensure the existence of this second extension. It also uses a characterization of idempotents in the algebra $B(G)$, namely, idempotents are characteristic functions of subsets in the coset ring $\Omega(G)$. This characterization was found by Cohen [3] for abelian groups, and by Host [14] for general groups. Interestingly, Host's proof is simpler than Cohen's.

The theorem of Cohen cannot extend to arbitrary locally compact groups without some restrictions. Indeed, take a free set F in a non-abelian free group G, and define $\phi \colon G \to G$ to be the identity off F, and to permute F in an arbitrary way. Then the associated map α is a complete isomorphism of $A(G)$, but ϕ is not nearly extended affine unless F is finite. Thus Cohen's theorem does not hold in this case. However, it is still possible that if $\phi \colon A(G') \to A(G)$ is an isomorphism, then G and G' are isomorphic by a map other than the map α arising from ϕ. Further, if we assume that α extends to a bounded map of $B(G')$ to $B(G)$, then the free set F must be finite and ϕ is nearly affine.

For connected groups, nearly affine maps are automatically affine, so the case where the groups are connected should be simpler. According to the structure theory of locally compact groups [20], connected locally compact groups are almost Lie groups; more precisely, any connected locally compact group G has a normal subgroup N contained in an arbitrarily small neighbourhood of the identity such that G/N is a Lie group.

1.3. The connection between cosets and function algebras

The following theorem relates cosets and function algebras on a locally compact group G.

Theorem 1.6. *A closed subset S of G is a coset of an amenable subgroup if and only if there exists a net $(u_i)_{i\in I}$ in $A_p(G)$ such that*

1. $\|u_i\|_{A_p} \le 1$ *for all* $i \in I$
2. $\lim_{i \in I} u_i(x) = 1$ *for all* $x \in S$
3. $\lim_{i \in I} u_i(x) = 0$ *for all* $x \in G \setminus S$.

Proof. The "if" part of the characterization holds because, by a theorem of M.G. Cowling and G. Fendler [5], the pointwise limit of a sequence or net of matrix coefficients of a representation is still a matrix coefficient of a representation (which may not be continuous). More precisely, there exists a representation π of G_d (the group G with the discrete topology) on a Banach space X, which is a subquotient of an L^p space, and unit vectors ξ and η in X, such that

$$\langle \pi(\cdot)\xi, \eta \rangle = \chi_S,$$

where χ_S denotes the characteristic function of S. Then

$$\langle \pi(\cdot)\xi, \pi(z^{-1})^*\eta \rangle = \chi_{zS};$$

we choose z so that $e \in zS$.

Since X is a subquotient of an L^p space, both X and X^* are strictly convex, and if we hold ξ fixed, then the unique unit vector ζ in X^* such that $\langle \xi, \pi(z^{-1})^*\zeta \rangle = 1$ is η, and if we hold η fixed, then the unique unit vector θ in X such that $\langle \theta, \pi(z^{-1})^*\eta \rangle = 1$ is ξ. It follows that zS, the set of x in G such that $\langle \pi(x)\xi, \pi(z^{-1})^*\eta \rangle = 1$, is the set of G of elements that fix $\pi(z^{-1})\xi$, which is a subgroup, and so S is a coset.

Conversely, by a theorem of C.S. Herz [13], for any closed subgroup H of G, the space of restrictions of $A_p(G)$-functions to H, equipped with the quotient norm, is exactly $A_p(H)$. Herz's proof shows that the extensions of $A_p(H)$-functions to $A_p(G)$ can have supports arbitrarily close to H. Since $A_p(H)$ has an approximate identity, we can use Herz's construction to produce a net $(u_i)_{i\in I}$ with the required properties. $\quad\square$

A similar theorem may be found in [21]. In conclusion, any isometric isomorphism of algebras $A_p(G_1)$ and $A_p(G_2)$ induces a homeomorphism of G_1 and G_2 that maps cosets of closed amenable subgroups to cosets of closed amenable subgroups.

2. Coset geometry of groups

In this section, we review some results about maps of groups that preserve cosets, in the sense that the image of every coset is a coset (more precisely, a possibly different coset of a possibly different subgroup). The first result is often called "the fundamental theorem of affine geometry".

Theorem 2.1. *Every homeomorphism of the plane* $\mathbb{R}^2$ *that sends lines to lines is composed of a translation and a linear map.*

The next result is new (to the best of my knowledge). It (and its extension to three-dimensional groups) will enable us to get a hold on maps that preserve cosets of more general groups.

Theorem 2.2. *Suppose that* G *and* G' *are two-dimensional Lie groups, and that* $\phi \colon G \to G'$ *is a homeomorphism. Suppose also that* S *is a coset of a closed subgroup of* G *if and only if* $\phi(S)$ *is a coset of a closed subgroup of* G'*. Then* G *and* G' *are isomorphic, and* ϕ *is extended affine. In particular,* ϕ *is a smooth diffeomorphism.*

Proof. The only two-dimensional Lie groups that are not simply connected are $\mathbb{R} \times \mathbb{T}$, the direct product of a line and a torus, and $\mathbb{T}^2$, the product of two tori. There are two simply connected Lie groups of dimension two: $\mathbb{R}^2$ and the "$ax + b$ group". Thus if G and G' are homeomorphic, and not simply connected, then they are also isomorphic. To deal with the groups that are not simply connected is not very difficult, and we omit many of the details here.

Henceforth, we assume that G and G' are simply connected. The cosets of closed connected one-dimensional subgroups of $\mathbb{R}^2$ are straight lines. By the fundamental theorem of affine geometry, the group CosetMaps($\mathbb{R}^2$) of all homeomorphisms of $\mathbb{R}^2$ that send these cosets to cosets is $\mathrm{GL}(2, \mathbb{R}) \ltimes \mathbb{R}^2$, which coincides with ExtAff($\mathbb{R}^2$).

Suppose that H is the "$ax + b$ group", consisting of all matrices of the form

$$\begin{pmatrix} a & b \\ 0 & 1 \end{pmatrix},$$

where $a \in \mathbb{R}^+$ and $b \in \mathbb{R}$. The cosets of closed connected one-dimensional subgroups in H are all of the form

$$\left\{ \exp\left(t \begin{pmatrix} x & y \\ 0 & 0 \end{pmatrix} \right) \begin{pmatrix} a & b \\ 0 & 1 \end{pmatrix} : t \in \mathbb{R} \right\},$$

where $(x, y) \in \mathbb{R}^2 \setminus \{0\}$. Computation shows that, if $x = 0$, then these are the sets

$$\left\{ \begin{pmatrix} a & b + ty \\ 0 & 1 \end{pmatrix} : t \in \mathbb{R} \right\},$$

while, when $x \neq 0$, these are the sets

$$\left\{ \begin{pmatrix} as & (b + y/x)s - y/x \\ 0 & 1 \end{pmatrix} : s \in \mathbb{R}^+ \right\}$$

(one writes e^{tx} as s). These may be identified with the intersections of straight lines in $\mathbb{R}^2$ with the right half plane $\mathbb{R}^2_+$ in $\mathbb{R}^2$. By a result of Čap, Cowling, De Mari, Eastwood and McCallum [2], the group CosetMaps(H) of homeomorphisms of H preserving cosets of one-dimensional subgroups "is" the group of homeomorphisms of $\mathbb{R}^2_+$ that send intersections of lines with $\mathbb{R}^2_+$ into intersections of lines with $\mathbb{R}^2_+$. This is the group of linear affine maps of $\mathbb{R}^2_+$ that preserve the y axis, extended by a projective map that exchanges the projective line at infinity with the y axis. This coincides with the group ExtAff(H).

If there were any coset-preserving homeomorphisms from $\mathbb{R}^2$ to H, they would intertwine the actions of the groups CosetMaps($\mathbb{R}^2$) and CosetMaps(H). The groups CosetMaps($\mathbb{R}^2$) and CosetMaps(H) are different, so there are no coset-preserving homeomorphisms from $\mathbb{R}^2$ to H. Further, when G and G' are isomorphic and coincide with one of $\mathbb{R}^2$ and H, than all coset-preserving homeomorphisms are extended affine. $\square$

Theorem 2.3. *Suppose that G and G' are three-dimensional Lie groups, and that $\phi \colon G \to G'$ is a homeomorphism. Suppose also that S is a coset of a closed subgroup of G if and only if $\phi(S)$ is a coset of a closed subgroup of G'. Then G and G' are isomorphic, and ϕ is extended affine. In particular, ϕ is a smooth diffeomorphism.*

Proof. The three-dimensional case is more complex than the two-dimensional case, as there are more possibilities; we shall not give a complete proof here. There are various simply connected Lie groups of dimension three (up to isomorphism), including $\mathbb{R}^3$, the Heisenberg group H^1, which is the only nonabelian nilpotent Lie group of dimension 3, several solvable groups, the unitary group SU(2), and the noncompact simple Lie group SL(2, $\mathbb{R}$).

The proof relies on studying each of these cases, examining the coset structure, showing that there cannot be coset-preserving and subgroup-preserving maps between the different cases, and showing the the coset-preserving and subgroup-preserving maps between the different cases corresponding to the extended affine maps. We give the details of one of these in the next section. $\square$

It should be pointed out that there are some quite old results that state that homeomorphic compact Lie groups are locally isomorphic (see, for instance, [15, 22]). The extra information that we use to show that our homeomorphism is an isomorphism is knowledge of the cosets. When these groups have abelian subgroups of dimension two, it is the fundamental theorem of affine geometry that does this. In the rank one case, closed one-dimensional cosets in SU(2) are known to correspond with great circles on the sphere S^3, and great-circle-preserving maps of spheres are already understood (see, for instance, [17]). Not all great-circle-preserving maps of the map S^3 correspond to extended affine maps of SU(2); we need to consider cosets of finite subgroups to prove the full result. Similarly, it is necessary to consider cosets of discrete subgroups to deal with the Heisenberg group.

3. An analysis of cosets and coset-preserving maps

We define the map $\tau : \mathbb{R} \to GL(2, \mathbb{R})$ by

$$\tau(t) = e^{\alpha t} \begin{pmatrix} \cos \beta t & \sin \beta t \\ -\sin \beta t & \cos \beta t \end{pmatrix} \quad \forall t \in \mathbb{R},$$

where α and β are nonzero real parameters, and define the group S to be the semidirect product $\mathbb{R} \ltimes \mathbb{R}^2$, where τ is the action of $\mathbb{R}$ on $\mathbb{R}^2$. Thus a typical element of S may be written as (s, u), where $s \in \mathbb{R}$ and $u \in \mathbb{R}^2$; further, $(s, u)^{-1} = (-s, -\tau(-s)u)$ and

$$(s, u)(t, v) = (s + t, \tau(t)u + v) \quad \forall s, t \in \mathbb{R} \quad \forall u, v \in \mathbb{R}^2.$$

Without loss of generality, we may assume that $\alpha > 0$; otherwise we just reparametrize the $\mathbb{R}$ factor in S, changing t to $-t$.

Lemma 3.1. *Suppose that S is the semidirect product just defined. The nontrivial connected subgroups of S are of one of the following three forms:*

(A) $\{(0, sv) : s \in \mathbb{R}\}$, *where $v \in \mathbb{R}^2 \setminus \{0\}$;*
(B) $\{(0, u) : u \in \mathbb{R}^2\}$;
(C) $\{(s, \tau(s)v - v : s \in \mathbb{R}\}$, *where $v \in \mathbb{R}^2$.*

All subgroups of type (A) are conjugate, as are all subgroups of type (C).

Proof. Connected subgroups of S correspond to subalgebras of the Lie algebra $\mathfrak{s}$ of S. The Lie algebra $\mathfrak{s}$ is spanned by T, X and Y, where $[X, Y] = 0$ and

$$[T, \lambda X + \mu Y] = (\lambda \alpha + \mu \beta)X + (\mu \alpha - \lambda \beta)Y \quad \forall \lambda, \mu \in \mathbb{R}.$$

If a subalgebra contains nonzero elements of the form $T + \xi X + \eta Y$ and $\lambda X + \mu Y$, then it also contains $(\lambda \alpha + \mu \beta) X + (\mu \alpha - \lambda \beta) Y$; the latter two elements are linearly independent so the subalgebra contains span$\{X, Y\}$, and hence is span$\{T, X, Y\}$. If a subalgebra contains two distinct elements of the form $T + \xi X + \eta Y$ and $T + \xi' X + \eta' Y$, then it contains a nonzero element of the form $\xi'' X + \eta'' Y$, and so is span$\{T, X, Y\}$ by the previous argument. Hence the only nontrivial subalgebras of $\mathfrak{s}$ are of the form span$\{U\}$, where $U \in \mathfrak{s} \setminus \{0\}$, and span$\{X, Y\}$.

The Lie subalgebras of type (A) are distinguished amongst the one-dimensional subalgebras as being those contained in span$\{X, Y\}$, the only two-dimensional subalgebra. The characterization of connected subgroups follows by passing this characterization of the subalgebras to the group.

Observe that

$$(s, 0)(0, tv)(s, 0)^{-1} = (s, tv)(-s, 0) = (0, \tau(-s)tv),$$

and the conjugacy of subgroups of type (A) follows. Further,

$$(s, u)(t, 0), (s, u)^{-1} = (s + t, \tau(t)u)(-s, -\tau(-s)u)$$
$$= (t, \tau(t)\tau(-s)u - \tau(-s)u)$$

and taking v to be $\tau(-s)u$, we have proved the conjugacy of subalgebras of type (C). $\qquad\square$

Lemma 3.2. *The nontrivial connected cosets of subgroups of the group S are all of one of the following forms:*

(A) $\{(t, sv + w) : s \in \mathbb{R}\}$, *where* $t \in \mathbb{R}$, *while* $v, w \in \mathbb{R}^2$ *and* $v \neq 0$;
(B) $\{(t, u) : u \in \mathbb{R}^2\}$, *where* $t \in \mathbb{R}$;
(C) $\{(s, \tau(s)v + w : s \in \mathbb{R}\}$, *where* $v, w \in \mathbb{R}^2$.

Given two distinct points p and q in S, either there is exactly one coset of type (A) that contains both p and q, or there are infinitely many cosets of type (C) that contain both.

Proof. This is an easy extension of the previous lemma. $\qquad\square$

Lemma 3.3. *The automorphism group* $\mathrm{Aut}(S)$ *consists of all maps* $\phi : S \to S$ *of the form*

$$\phi(t, u) = (t, \tau(t)v - v + Lu) \quad \forall t \in \mathbb{R} \quad \forall u \in \mathbb{R}^2, \qquad (3.1)$$

where $v \in \mathbb{R}^2$ *and* L *in* $\mathrm{SL}(2, \mathbb{R})$ *is the composition of a dilation and a rotation.*

Proof. First, if ϕ is an automorphism, then ϕ must map $\{0\} \times \mathbb{R}^2$ to itself, sending the origin of $\mathbb{R}^2$ to itself, and cosets in this subgroup into cosets, that is, sending lines in $\mathbb{R}^2$ into lines in $\mathbb{R}^2$. By the fundamental theorem of affine geometry, there is a linear map L such that

$$\phi(0, u) = (0, Lu) \quad \forall u \in \mathbb{R}^2.$$

Next, ϕ must map one-parameter subgroups of type (C) to one-parameter subgroups of type (C), whence there exist $\gamma \in \mathbb{R} \setminus \{0\}$ and v in $\mathbb{R}^2$ such that

$$\phi(s, 0) = (\gamma s, \tau(\gamma s)v - v) \quad \forall s \in \mathbb{R}.$$

Since ϕ is an automorphism,

$$\phi(s, u) = (s, \tau(\gamma s)v - v + Lu) \quad \forall s \in \mathbb{R} \quad \forall u \in \mathbb{R}^2.$$

It remains to show that $\gamma = 1$ and that L is composed of a dilation and a rotation. Observe that

$$\phi((s, 0)(0, u)(s, 0)^{-1}) = \phi(s, 0)\phi(0, u)\phi(s, 0)^{-1} \quad \forall s \in \mathbb{R} \quad \forall u \in \mathbb{R}^2,$$

and hence

$$\begin{aligned}
(0, L\tau(-s)u) &= \phi(0, \tau(-s)u) \\
&= \phi(s, 0)\phi(0, u)\phi(s, 0)^{-1} \\
&= (\gamma s, \tau(\gamma s)v - v)(0, Lu)(-\gamma s, -v + \tau(-\gamma s)v) \\
&= (0, \tau(-\gamma s)Lu) \quad \forall s \in \mathbb{R} \quad \forall u \in \mathbb{R}^2.
\end{aligned}$$

By considering determinants, we see that

$$\det(L)\det(\tau(-s)) = \det(\tau(-\gamma s))\det(L),$$

whence $\gamma = 1$, and $L\tau(s) = \tau(s)L$ for all s in $\mathbb{R}$; this implies that L is of the stated form.

Conversely, it is easy to check that ϕ is an automorphism if it is of the form (3.1). $\qquad\square$

Finally, to prove the main theorem in this section, we will need to solve some matrix equations, and the following lemma will be helpful.

Lemma 3.4. *Suppose that $\delta \colon \mathbb{R} \to \mathbb{R}$ is a bijection and $\delta(0) = 0$. Suppose also that $M_1, M_2 \in \mathrm{GL}(2, \mathbb{R})$ and M_3 is a 2×2 real matrix. If*

$$\tau(\delta(s))M_1 = M_2\tau(s) + M_3 \quad \forall s \in \mathbb{R}, \tag{3.2}$$

then $\delta(s) = s$, while $M_1 = M_2$ and $M_3 = 0$; further, M_1 is composed of a rotation and a dilation.

Proof. The right hand side of (3.2) varies smoothly with s, so δ is a smooth function. Smooth bijections are either increasing or decreasing, and consideration of the determinants of both sides of (3.2) shows that δ must be increasing. Hence, sending s to $-\infty$, we deduce that $M_3 = 0$, and since $\delta(0) = 0$, we deduce that $M_1 = M_2$. Thus

$$\tau(\delta(s))M_1 = M_1\tau(s) \quad \forall s \in \mathbb{R}.$$

Taking determinants shows that $\delta(s) = s$; since M_1 commutes with $\tau(s)$ for all $s \in \mathbb{R}$, it follows that M_1 is composed of a dilation and a rotation. $\qquad\qquad$ \ $\square$

Theorem 3.5. *Suppose that $\phi\colon S \to S$ is a homeomorphism, and that the image of each coset of a connected subgroup of S is a coset of a connected subgroup. Then ϕ is composed of a translation, an automorphism, and possibly inversion.*

Proof. By composing with a translation, we may suppose that $\phi(0, 0) = (0, 0)$; as ϕ maps cosets of connected two-dimensional subgroups to cosets of connected two-dimensional subgroups, ϕ maps $\{0\} \times \mathbb{R}^2$ onto itself. More generally, ϕ maps $\{s\} \times \mathbb{R}^2$ onto $\{\gamma(s)\} \times \mathbb{R}^2$, for some homeomorphism $\gamma\colon \mathbb{R} \to \mathbb{R}$. Every such homeomorphism is either increasing or decreasing; by composing with the group inverse if necessary, we may suppose that γ is increasing.

Next, ϕ maps $\mathbb{R} \times \{0\}$ to a subgroup of type (C), and by composing with an automorphism we may suppose that ϕ maps $\mathbb{R} \times \{0\}$ to itself. Finally, ϕ maps type (A) cosets in $\{s\} \times \mathbb{R}^2$ to type (A) cosets in $\{\gamma(s)\} \times \mathbb{R}^2$ and maps $(s, 0)$ to $(\gamma(s), 0)$, so by the fundamental theorem of affine geometry, there exists a linear bijection $L(s)\colon \mathbb{R}^2 \to \mathbb{R}^2$, possibly depending on s, such that

$$\phi(s, u) = (\gamma(s), L(s)u) \quad \forall s \in \mathbb{R} \quad \forall u \in \mathbb{R}^2.$$

Clearly ϕ maps type (C) cosets to type (C) cosets, so for all $u, v \in \mathbb{R}^2$, there exist $\tilde{u}, \tilde{v} \in \mathbb{R}^2$ such that

$$\phi(s, \tau(s)u + v) = (\gamma(s), \tau(\gamma(s))\tilde{u} + \tilde{v}) \quad \forall s \in \mathbb{R}.$$

This implies that

$$\tau(\gamma(s))\tilde{u} + \tilde{v} = L(s)(\tau(s)u + v)$$

for all $u, v \in \mathbb{R}^2$ and $s \in \mathbb{R}$. In particular, taking s to be 0, we see that

$$\tilde{u} + \tilde{v} = L(0)(u + v),$$

so

$$(\tau(\gamma(s)) - I)\tilde{u} = (L(s)\tau(s) - L(0))u + (L(s) - L(0))v$$
$$(\tau(\gamma(s)) - I)\tilde{v} = (\tau(\gamma(s))L(0) - L(s)\tau(s))u$$
$$+ (\tau(\gamma(s))L(0) - L(s))v.$$

This implies that there are linear maps A, B, C and D of $\mathbb{R}^2$ such that

$$\begin{pmatrix} \tilde{u} \\ \tilde{v} \end{pmatrix} = \begin{pmatrix} A & B \\ C & D \end{pmatrix} \begin{pmatrix} u \\ v \end{pmatrix}.$$

These maps are independent of s in $\mathbb{R}$, and so for all $s \in \mathbb{R} \setminus \{0\}$,

$$A = (\tau(\gamma(s)) - I)^{-1}(L(s)\tau(s) - L(0))$$
$$B = (\tau(\gamma(s)) - I)^{-1}(L(s) - L(0))$$
$$C = (\tau(\gamma(s)) - I)^{-1}(\tau(\gamma(s))L(0) - L(s)\tau(s))$$
$$D = (\tau(\gamma(s)) - I)^{-1}(\tau(\gamma(s))L(0) - L(s)),$$

or equivalently,

$$(\tau(\gamma(s)) - I)A = L(s)\tau(s) - L(0) \tag{3.3}$$
$$(\tau(\gamma(s)) - I)B = L(s) - L(0) \tag{3.4}$$
$$(\tau(\gamma(s)) - I)C = \tau(\gamma(s))L(0) - L(s)\tau(s)$$
$$(\tau(\gamma(s)) - I)D = \tau(\gamma(s))L(0) - L(s).$$

Since γ is increasing, it is unbounded in $\mathbb{R}^+$. Take u in $\mathbb{R}^2$ such that $\|u\| = 1$ and $\|Bu\| = \|B\|$. From (3.4),

$$\|(\tau(\gamma(s)) - I)Bu\| = \|(L(s) - L(0))u\|,$$

so

$$\|L(s)u\| \geq \|\tau(\gamma(s))Bu\| - \|Bu\| - \|L(0)u\|$$
$$\geq e^{\alpha\gamma(s)}\|B\| - \|B\| - \|L(0)\|. \tag{3.5}$$

On the other hand, from (3.3),

$$(\tau(\gamma(s)) - I)A\tau(-s)u = (L(s)\tau(s) - L(0))\tau(-s)u,$$

whence

$$\|L(s)u\| \leq \|\tau(\gamma(s)A\tau(-s)u\| + \|A\tau(-s)u\| + \|L(0)\tau(-s)u\|$$
$$\leq e^{\alpha[\gamma(s)-s]}\|A\| + e^{-\alpha s}[\|A\| + \|L(0)\|],$$

and combined with (3.5), this shows that $B = 0$. It follows that $L(s) = L(0)$ for all $s \in \mathbb{R}$, and from (3.3),

$$\tau(\gamma(s))A = L(0)\tau(s) + [A - L(0)].$$

By Lemma 3.4, $\gamma(s) = s$, while $A = L(0)$; further, $L(0)$ is composed of a dilation and a rotation. We also see that $C = 0$ and $D = L(0)$. Thus

$$\phi(s, u) = (\gamma(s), L(0)u) \quad \forall s \in \mathbb{R} \quad \forall u \in \mathbb{R}^2,$$

and ϕ is an automorphism. $\square$

4. Coset-preserving maps of connected Lie groups

In this section, we consider coset-preserving maps of connected Lie groups. We consider only cosets of closed subgroups. Our aim is to establish smoothness and some Lie algebraic structure of these maps.

Theorem 4.1. *Suppose that G and G' are connected Lie groups, and that $\phi : G \to G'$ is a homeomorphism that sends cosets of closed amenable subgroups to cosets of closed amenable subgroups. Then ϕ is smooth, and after composition with a translation, sends the identity of G to the identity of G'. The derivative $d\phi : \mathfrak{g} \to \mathfrak{g}'$ of ϕ at the identity is a linear map of the Lie algebra, and satisfies*

- $\exp(d\phi(X)) = \phi(\exp(X))$ *for all $X \in \mathfrak{g}$*
- $d\phi[X, [X, Y]] = [d\phi(X), [d\phi(X), d\phi(Y)]]$ *for all $X, Y \in \mathfrak{g}$*
- $d\phi$ *sends subalgebras corresponding to closed amenable subgroups to subalgebras corresponding to closed amenable subgroups.*

Proof. By composing with a translation if necessary, we may suppose for the moment that $\phi(e) = e$. Let $X \in \mathfrak{g}$, and consider $\exp(\mathbb{R}X)$.

If $\exp(\mathbb{R}X)$ is closed, then the restriction of ϕ to $\exp(\mathbb{R}X)$ maps onto a one-paramenter subgroup $\exp(\mathbb{R}Y)$, and induces a 0-preserving homeomorphism $\tilde{\phi}$ of $\mathbb{R}$ that sends each coset $a + b\mathbb{Z}$ to a coset $c + d\mathbb{Z}$; using the fact that homeomorphisms of $\mathbb{R}$ are monotone, it is easy to see that $\tilde{\phi}$ is linear, and hence ϕ is an isomorphism.

If $\exp(\mathbb{R}X)$ is not closed, its closure A is a closed compact abelian subgroup of G, and the restriction of ϕ to A is a coset-preserving homeomorphism. It is not hard to see that the restricted map is an isomorphism. In either case, it follows that $\exp(d\phi(tX)) = \phi(t\exp(X))$ for all $t \in \mathbb{R}$.

We now relax the condition that $\phi(e) = e$, and deduce that ϕ is affine and hence smooth on cosets of one-parameter subgroups of G. A map that is smooth on all cosets of all one-parameter subgroups is smooth.

Further, at least locally, ϕ preserves the symmetry of reflection in any point, that is, $\phi(xy^{-1}x) = \phi(x)\phi(y)^{-1}\phi(x)$, at least if x and y are close. The infinitesimal version of this condition is that $d\phi$ is Jordan affine, that is,

$$d\phi[X, [X, Y]] = [d\phi(X), [d\phi(X), d\phi(Y)]] \quad \forall X, Y \in \mathfrak{g},$$

as required. $\qquad\qquad\qquad\qquad\qquad\qquad\qquad\qquad\qquad\qquad\qquad\quad\square$

Note that the Jordan affine condition implies that $d\phi$ maps nilpotent elements of $\mathfrak{g}$ to nilpotent elements. In the semisimple case, this is enough to prove the result, as R. Guralnick [11] has shown that a linear map of a semisimple Lie algebra that sends nilpotent elements to nilpotent elements is an isomorphism or anti-isomorphism.

5. Isometries of the algebras $A_p(G)$

In this section, we put together what we know to prove our main theorem.

Theorem 5.1. *Suppose that G and G' are connected Lie groups, and $\alpha: A_p(G') \to A_p(G)$ is an isometric isomorphism. Then the associated homeomorphism $\phi: G \to G'$ is extended affine.*

Proof. If G and G' are one-dimensional, and $A_p(G)$ and $A_p(G')$ are isomorphic, then G and G' are abelian, and the induced map of groups is an isomorphism, by Lohoué's theorem (Theorem 1.3). In what follows, we need only deal with groups of dimension at least two, and in light of our discussion about groups of dimension two earlier, we may limit our attention to groups of dimension at least 3.

The previous theorem shows that ϕ is smooth. Further, by translating if necessary to ensure that $\phi(e) = e$, then for elements X of $\mathfrak{g}$ for which $\text{ad}(X)$ is diagonalisable with real eigenvalues, and corresponding eigenvectors Y,

$$d\phi[X, Y] = \pm[d\phi(X), d\phi(Y)],$$

which is stronger than the Jordan affine condition. This is enough to ensure that for many groups, such as generalisations of the $ax + b$ group where $a \in \mathbb{R}^+$ and $b \in \mathbb{R}^n$, the theorem holds. On the other hand, when G is nilpotent of step 2, the Jordan affine condition is vacuous, and no X in $\mathfrak{g}$ has eigenvectors. Thus the nilpotent case is in some sense the hardest part of the problem.

For brevity, we consider only the case where G is nilpotent and simply connected, and show how to treat this case. The general result then follows from stitching together this and similar information.

If G is nilpotent, then it is amenable as a discrete group, and we may use a recent result about $A_p(G)$ due to A. Derighetti [6]: the set of restrictions of $A_p(G)$ functions to a finite subset F of G agrees with the restriction to F of $A_p(G_d)$ functions. It follows that, if G and G' are amenable as discrete groups, and $A_p(G')$ and $A_p(G)$ are isometrically isomorphic, then $A_p(G'_d)$ and $A_p(G_d)$ are also isometrically isomorphic. We deduce that ϕ maps cosets of arbitrary (not necessarily closed) subgroups of G to G'.

In the nilpotent case, the Baker–Campbell–Hausdorff formula is easy to use, as there are no convergence issues. Recall that

$$\exp(X)\exp(Y) = \exp(\mathrm{BCH}(X, Y)) \quad \forall X, Y \in \mathfrak{g},$$

where BCH is a polynomial in two variables:

$$\mathrm{BCH}(X, Y) = X + Y + \frac{1}{2}[X, Y] + \dots.$$

Take a small positive real parameter t, and consider the group $\langle tX, tY\rangle$ generated by $\exp(tX)$ and $\exp(tY)$. From the Baker–Campbell–Hausdorff formula, the elements of this group vary smoothly with t, and the union $\bigcup_{t\in\mathbb{R}^+}\langle tX, tY\rangle$ is a union of one-dimensional submanifolds of G. Indeed, there exists an integer k such that we may represent all the elements of $\langle tX, tY\rangle$ in the form

$$\exp(m_1 tX)\exp(n_1 tY)\exp(m_2 tx)\exp(m_2 tY)\dots\exp(m_k tX)\exp(n_k tY),$$

where m_1, n_1, and so on are all integers. For every $2k$-tuple of integers, we obtain a curve in G:

$$\gamma(t) = \exp(m_1 tX)\exp(n_1 tY)\exp(m_2 tx)\exp(m_2 tY)\dots$$
$$\dots\exp(m_k tX)\exp(n_k tY).$$

We may divide these curves into equivalence classes. The first collection of equivalence classes is composed of curves such that

$$\log\gamma(t) = t(mX + nY) + o(t)$$

for some integers m and n, not both of which are zero. The next collection of equivalence classes is composed of curves such that

$$\log\gamma(t) = t^2 m[X, Y] + o(t^2),$$

for some nonzero integer m. Then there are curves such that

$$\log\gamma(t) = O(t^3).$$

From these curves and some differential calculus, we can recover the set $\{m[X, Y] : m \in \mathbb{Z}\}$; this set has two generators, namely $\pm[X, Y]$. Hence $d\phi[X, Y] = \pm[d\phi X, d\phi Y]$.

Now the sets

$$\{(X, Y) \in \mathfrak{g}^2 : d\phi[X, Y] = [d\phi(X), d\phi(Y)]\}$$

and

$$\{(X, Y) \in \mathfrak{g}^2 : d\phi[X, Y] = -[d\phi(X), d\phi(Y)]\}$$

are Zariski closed in $\mathfrak{g}^2$, and their union is $\mathfrak{g}^2$; then one of them must be the whole space, and so $d\phi$ is either a Lie algebra homomorphism or a Lie algebra anti-homomorphism. $\square$

6. A final remark

When $p \neq 2$, it seems likely that we can say a little more.

Conjecture 6.1. Suppose that G and G' are connected Lie groups, that $p \in (1, \infty) \setminus \{2\}$, and that $A_p(G)$ and $A_p(G')$ are isometrically isomorphic as Banach algebras. Then the map $\phi \colon G \to G'$ such that $\alpha u = u \circ \phi$ for all $u \in A_p(G')$ is an isomorphism.

This should follow from the following argument. First, we know that ϕ preserves subgroups and is an isomorphism or an anti-isomorphism; to decide between these possibilities, it is enough to restrict to a small nonabelian subgroup and decide whether the restricted map is an isomorphism or anti-isomorphism. Next, quite a lot of work has gone into studying small nonabelian subgroups, and we have almost enough information to be able to decide the question, due largely to work of A.M. Mantero [19] and of A.H. Dooley, S. Gupta and F. Ricci [7] (inspired by others before them). It would suffice to know that the conjecture was true for the "$ax + b$ group" to know it in general.

References

[1] M. BARONTI, *Algebre di Banach A_p di gruppi localmente compatti*, Riv. Mat. Univ. Parma **11** (1985), 399–407.

[2] A. ČAP, M. G. COWLING, F. DE MARI, M. G. EASTWOOD and R. G. MCCALLUM, *The Heisenberg group,* SL(3, **R**) *and rigidity*, pages 41–52, In: "Harmonic Analysis, Group Representations, Automorphic Forms and Invariant Theory", edited by Jian-Shu Li, Eng-Chye Tan, Nolan Wallach and Chen-Bo Zhu. World Scientific, Singapore, 2007.

[3] P. M. COHEN, *On a conjecture of Littlewood and idempotent measures*, Amer. J. Math. **82** (1960), 191–212.

[4] P. M. COHEN, *On homomorphisms of group algebras*, Amer. J. Math. **82** (1960), 213–226.

[5] M. G. COWLING and G. FENDLER, *On representations in Banach spaces*, Math. Ann. **266** (1984), 307–315.

[6] A. DERIGHETTI, *A property of* $B_p(G)$. *Applications to convolution operators*, J. Funct. Anal. **256** (2009), 928–939.

[7] A. H. DOOLEY, S. GUPTA and F. RICCI, *Asymmetry of convolution norms on Lie groups*, J. Funct. Anal. **174** (2000), 399–416.

[8] P. EYMARD, *L'algèbre de Fourier d'un groupe localement compact*, Bull. Soc. Math. France **92** (1964), 181–236.

[9] P. EYMARD, *Algèbres* A_p *et convoluteurs de* L^p, pages 55–72 (exposé 367) In: "Séminaire Bourbaki", Vol. 1969/1970, Exposés 364–381, Springer-Verlag, Berlin, Heidelberg, New York, 1971.

[10] A. FIGÀ-TALAMANCA, *Translation invariant operators in* L^p, Duke Math. J. **32** (1965), 495–501.

[11] R. GURALNICK, *Invertibility preservers and algebraic groups*, Linear Algebra Appl. **212/213** (1994), 249–257.

[12] C. S. HERZ, *Remarques sur la note précédente de M. Varopoulos*, C. R. Acad. Sci. Paris **260** (1965), 6001–6004.

[13] C. S. HERZ, *Harmonic synthesis for subgroups*, Ann. Inst. Fourier (Grenoble) **23**(3) (1973), 91–123.

[14] B. HOST, *Le théorème des idempotents dans* $B(G)$, Bull. Soc. Math. France **114** (1986), 215–221.

[15] J. R. HUBBUCK and R. M. KANE, *The homotopy types of compact Lie groups*, Israel J. Math. **51** (1985), 20–26.

[16] M. ILIE and N. SPRONK, *Completely bounded homomorphisms of the Fourier algebras*, J. Funct. Anal. **225** (2005), 480–499.

[17] J. JEFFERS, *Lost theorems of geometry*, Amer. Math. Monthly **107** (2000), 800–812.

[18] N. LOHOUÉ, *Sur certaines propriétés remarquables des algèbres* $A_p(G)$, C. R. Acad. Sci. Paris Sér. A-B **273** (1971), A893–A896.

[19] A. M. MANTERO, *Asymmetry of convolution operators on the Heisenberg group*, Boll. Un. Mat. Ital. A (6) **4** (1985), 19–27.

[20] D. MONTGOMERY and L. ZIPPIN, "Topological Transformation Groups", Interscience Publishers, New York-London, 1955.

[21] V. RUNDE, *Cohen–Host type idempotent theorems for representations on Banach spaces and applications to Figà-Talamanca–Herz algebras*, J. Math. Anal. Appl. **329** (2007), 736–751.

[22] H. SCHEERER, *Homotopieäquivalente kompakte Liesche gruppen*, Topology **7** (1968), 227–232.

[23] M. E. WALTER, *A duality between locally compact groups and certain Banach algebras*, J. Funct. Anal. **17** (1974), 131–160.

Classical analysis and nilpotent Lie groups

Joseph A. Wolf

Classical Fourier analysis has an exact counterpart in group theory and in some areas of geometry. Here I'll describe how this goes for nilpotent Lie groups and for a class of Riemannian manifolds closely related to a nilpotent Lie group structure. There are also some infinite dimensional analogs but I won't go into that here. The analytic ideas are not so different from the classical Fourier transform and Fourier inversion theories in one real variable.

Here I'll give a few brief indications of this beautiful topic. References, proofs and related topics for the finite dimensional theory can be found a recent AMS Monograph/Survey volume [1]. If you are interested in the infinite dimensional theory you may also wish to look at the article [2] in Mathematische Annalen.

In Section 1 I'll recall a few basic facts on classical Fourier theory and note the connection with the theory of locally compact abelian groups and their unitary representations. In Section 2 we look at the first non-commutative locally compact groups, the Heisenberg groups H_n. We describe their unitary representations, Fourier transform theory and Fourier inversion formula.

The coadjoint orbit picture is the best way to understand representations of nilpotent Lie groups. It is guided by the example of the Heisenberg group. We indicate that theory in Section 3. Then in Section 4 we come to a class of connected, simply connected, nilpotent Lie groups with many of the good analytic properties of vector groups and Heisenberg groups. Those are the simply connected, nilpotent Lie groups with square integrable representations.

In Section 5 we push some of the analysis to a class of homogeneous spaces where the techniques and results are analogous to those of locally compact abelian groups. Those are the commutative spaces G/K, i.e. the Gelfand pairs (G, K). We have already had a glimpse of this in Section 2 for the semidirect products $H_n \rtimes K$ and the riemannian homogeneous spaces $(H_n \rtimes K)/K$ where K is a compact group of automorphisms of H_n. In Section 6 we look more generally at Fourier transform theory and the Fourier inversion formula for commutative nilmanifolds $(N \rtimes K)/K$ where N is a simply connected, nilpotent Lie groups with square integrable representations and K is a compact group of automorphisms of N.

As indicated earlier, Fourier analysis for nilpotent Lie groups N and commutative nilmanifolds $(N \rtimes K)/K$, where N has square integrable representations, has recently been extended to some classes direct limit groups and spaces.

1. Classical Fourier series

Let's recall the Fourier series development for a function f of one variable that is periodic of period 2π. One views f as a function on the circle $S = \{e^{ix}\}$. The circle S is a multiplicative group and we expand f in terms of the unitary characters

$$\chi_n : S \to S \text{ by } \chi_n(x) = e^{inx} \text{ , continuous group homomorphism.}$$

Then the Fourier inversion formula is

$$f(x) = \sum_{n=-\infty}^{\infty} \widehat{f}(n)\chi_n$$

where the Fourier transform

$$\widehat{f}(n) = \tfrac{1}{2\pi} \int_0^{2\pi} f(x)e^{-inx}dx = \langle f, \chi_n \rangle_{L^2(S)}.$$

The point is that f is a linear combination of the χ_n with coefficients given by the Fourier transform $\widehat{f}$. This uses the topological group structure and the rotation–invariant measure on S.

One has a similar situation when the compact group S is replaced by a finite dimensional real vector space V. Let V^* denote its linear dual space. If $f \in L^1(V) \cap L^2(V)$ the Fourier inversion formula is

$$f(x) = \left(\tfrac{1}{2\pi}\right)^{m/2} \int_{V^*} \widehat{f}(\xi)e^{ix\cdot\xi}d\xi$$

where the Fourier transform is

$$\widehat{f}(\xi) = \left(\tfrac{1}{2\pi}\right)^{m/2} \int_V f(x)e^{-ix\cdot\xi}dx = \langle f, \chi_\xi \rangle_{L^2(V)}.$$

Again, f is a linear combination (this time it is a continuous linear combination) of the unitary characters $\chi_\xi(x) = e^{ix\cdot\xi}$ on V, and the coefficients of the linear combination are given by the Fourier transform $\widehat{f}$.

It is the same story for locally compact abelian groups G. The unitary characters form a group

$$\widehat{G} = \{\chi : G \to S \text{ continuous homomorphisms}\}$$

with composition $(\chi_1\chi_2)(x) = \chi_1(x)\chi_2(x)$. It is locally compact with the weak topology for the evaluation maps $ev_x : \chi \mapsto \chi(x)$. If $f \in L^1(G) \cap L^2(G)$ the Fourier inversion formula is

$$f(x) = \int_{\widehat{G}} \widehat{f}(\chi)\chi(x)d\chi$$

where the Fourier transform

$$\widehat{f}(\chi) = \int_G f(x)\overline{\chi(x)}dx = \langle f, \chi \rangle_{L^2(G)}.$$

As in the Euclidean cases, Fourier inversion expresses the function f as a (possibly continuous) linear combination of unitary characters on G, where the coefficients of the linear combination are given by the Fourier transform.

In this context, $f \mapsto \widehat{f}$ preserves L^2 norm and extends by continuity to an isometry of $L^2(G)$ onto $L^2(\widehat{G})$. In effect this expresses $L^2(G)$ as a (possibly continuous) sum of G–modules,

$$L^2(G) = \int_{\widehat{G}} \mathbb{C}_\chi d\chi \text{ where } \mathbb{C}_\chi \text{ is spanned by } \chi.$$

In this direct integral decomposition $d\chi$ could be replaced by any equivalent measure, so that decomposition is not as precise as the Fourier inversion formula.

2. The Heisenberg group

Next, we see what happens when we weaken the commutativity condition. The first case of that is the case of the Heisenberg group. There the Fourier transform and Fourier inversion are in some sense the same as in the classical case of a vector group, except that some of the integration occurs in the character formula and the rest in integration over the unitary dual.

The Heisenberg group of real dimension $2m + 1$ is

$$H_m = \operatorname{Im} \mathbb{C} + \mathbb{C}^m \text{ with group law } (z, w)(z', w')$$
$$= (z + z' + \operatorname{Im} \langle w, w' \rangle, w + w')$$

where Im denotes imaginary component (as opposed to the coefficient of $\sqrt{-1}$), $z, z' \in Z := \operatorname{Im} \mathbb{C}$ and $w, w' \in W := \mathbb{C}^m$. Its Lie algebra, the Heisenberg algebra, is

$$\mathfrak{h}_m = \mathfrak{z} + \mathfrak{w} = \operatorname{Im} \mathbb{C} + \mathbb{C}^m \text{ with } [z + w, z' + w']$$
$$= (z + z' + 2 \operatorname{Im} \langle w, w' \rangle).$$

Here $Z = \exp(\mathfrak{z})$ is both the center and the derived group of H_m, and its complement $W = \exp(\mathfrak{w}) \cong \mathbb{R}^{2m}$.

Unitary characters have to annihilate the derived group of H_m, in other words factor through H_m/Z, so the only functions that can be expanded in unitary characters are the functions that are lifted from H_m/Z. Thus we have to consider something more general.

The space $\widehat{H_m}$ of (equivalence classes of) irreducible unitary representations of H_m breaks into two parts, one consisting of the 1–dimensional representations and the other of the infinite dimensional representation. This goes as follows.

- One-dimensional representations. These are the ones that annihilate the center Z, and are given by the unitary characters χ_ξ, $\xi \in W^*$, on $W \cong \mathbb{R}^{2m}$.
- Infinite dimensional representations. These are the $\pi_\zeta = \operatorname{Ind}_N^{H_m}(\chi_\zeta)$ where

$$N = \operatorname{Im} \mathbb{C} + \mathbb{R}^m \subset H_m \text{ and } \zeta \in \mathfrak{z}^* \setminus \{0\}.$$

Recall the definition of the induced representation $\pi_\zeta = \operatorname{Ind}_N^{H_m}(\chi_\zeta)$. Here χ_ζ extends from Z to N by $\chi_\zeta(z, w) = \chi_\zeta(z)$. Thus we have a unitary line bundle over H_m/N associated to the principal N–bundle $H_m \to H_m/N$ by the action $w : t \mapsto \chi_\zeta(w)t$ of N on $\mathbb{C}$. Now π_ζ is the natural action of H_m on the space of L^2 sections of that line bundle.

The classical "Uniqueness of the Heisenberg commutation relations" says that ζ determines the class $[\pi_\zeta] \in \widehat{H_m}$. And restriction to Z shows that $[\pi_\zeta] = [\pi_{\zeta'}]$ just when $\zeta = \zeta'$.

Using the fact that ζ determines $[\pi_\zeta]$, one realizes $[\pi_\zeta]$ by an action of H_m on the Hilbert space $\mathcal{H}_m$ of Hermite polynomials on $\mathbb{C}^m$. The maximal compact subgroup of $\operatorname{Aut}(H_m)$ is the unitary group $U(m)$. Its action is

$$g : (z, w) \mapsto (z, g(w)).$$

Result: π_ζ extends to an irreducible unitary representation $\widetilde{\pi}_\zeta$ of the semidirect product $H_m \rtimes U(m)$ on $\mathcal{H}_m$. So if K is any closed subgroup of $U(m)$ then $\widetilde{\pi}_\zeta|_{H_m \rtimes K}$ is an irreducible unitary representation of $H_m \rtimes K$ on $\mathcal{H}_m$. The Mackey Little Group theory says that $\widehat{H_m \rtimes K} = \{[\widetilde{\pi}_\zeta \otimes \kappa] \mid [\kappa] \in \widehat{K}$ and $\zeta \in \mathfrak{z}^* \setminus \{0\}\}$.

3. Representations and coadjoint orbits

Kirillov theory for connected simply connected nilpotent Lie groups N realizes their unitary representations in terms of the the coadjoint representation of N, that is, the representation ad^* of N on the linear dual space $\mathfrak{n}^*$ of its Lie algebra $\mathfrak{n}$.

On the group level the coadjoint representation is given by

$$(\mathrm{Ad}^*(n)f)(\xi) = f(\mathrm{Ad}(n)^{-1}\xi).$$

Write $\mathcal{O}_f$ for the (coadjoint) orbit $\mathrm{Ad}^*(N)f$ of the linear functional f. Consider the antisymmetric bilinear form b_f on $\mathfrak{n}^*$ given by $b_f(\xi, \eta) = f([\xi, \eta])$. The kernel of b_f is the Lie algebra of the isotropy subgroup of N at f. Thus b_f defines an $\mathrm{Ad}^*(N)$–invariant symplectic form ω_f on the coadjoint orbit $\mathcal{O}_f$. The symplectic homogeneous space $(\mathcal{O}_f, \omega_f)$ leads to a unitary representation class $[\pi_f] \in \widehat{N}$, as follows.

Let N_f denote the $\mathrm{Ad}^*(N)$–stabilizer of f. Its Lie algebra $\mathfrak{n}_f$ is the annihilator of f, in other words $\mathfrak{n}_f = \{v \in \mathfrak{n} \mid f(v, \mathfrak{n}) = 0\}$. A (real) **polarization** for f is a subalgebra $\mathfrak{p} \subset \mathfrak{n}$ that contains $\mathfrak{n}_f$, has dimension given by $\dim(\mathfrak{p}/\mathfrak{n}_f) = \frac{1}{2}\dim(\mathfrak{n}/\mathfrak{n}_f)$, and satisfies $f([\mathfrak{p}, \mathfrak{p}]) = 0$. Under the differential $\mathfrak{n} \to T_f(\mathcal{O}_f)$ of $N \to \mathcal{O}_f$, real polarizations for f are in one to one correspondence with N–invariant integrable Lagrangian distributions on $(\mathcal{O}_f, \omega_f)$.

Fix a real polarization $\mathfrak{p}$ for f and let $P = \exp(\mathfrak{p})$. It is the analytic subgroup of N for $\mathfrak{p}$ and it is a closed, connected, simply connected subgroup of N. In particular $e^{if} : P \to \mathbb{C}$ is a well defined unitary character. That defines a unitary representation

$$\pi_f = \pi_{f,\mathfrak{p}} = \mathrm{Ind}_P^N(e^{if})$$

of N. The basic facts are given by

Theorem 3.1. *Let N be a connected simply connected nilpotent Lie group and $f \in \mathfrak{n}^*$.*

1. *There exist real polarizations $\mathfrak{p}$ for f.*

2. *If $\mathfrak{p}$ is a real polarization for f then the unitary representation $\pi_{f,\mathfrak{p}}$ is irreducible.*

3. *If $\mathfrak{p}$ and $\mathfrak{p}'$ are real polarizations for f then the unitary representations $\pi_{f,\mathfrak{p}}$ and $\pi_{f,\mathfrak{p}'}$ are equivalent, so the class $[\pi_f] \in \widehat{N}$ is well defined.*

4. *If $[\pi] \in \widehat{N}$ then there exists $h \in \mathfrak{n}^*$ such that $[\pi] = [\pi_h]$.*

In other words, $f \mapsto \pi_{f,\mathfrak{p}}$ induces a one to one map of $\mathfrak{n}^*/\mathrm{Ad}^*(N)$ onto $\widehat{N}$.

To see just how this works, consider the case where N is the Heisenberg group H_m, and let $f \in \mathfrak{h}_m^*$. Here the center $\mathfrak{z} = \mathrm{Im}\,\mathbb{C}$ and its complement $\mathfrak{v} = \mathbb{C}^n$. Decompose $\mathfrak{v} = \mathfrak{u} + \mathfrak{w}$ where $\mathfrak{u} = \mathbb{R}^n$ and $\mathfrak{w} = i\mathbb{R}^n$. Note $\mathrm{Im}\,\langle \mathfrak{u}, \mathfrak{u}\rangle = 0 = \mathrm{Im}\,\langle \mathfrak{w}, \mathfrak{w}\rangle$. If $f(\mathfrak{z}) = 0$ we have the real polarization $\mathfrak{p} = \mathfrak{h}_m$. If $f(\mathfrak{z}) \neq 0$ we have the real polarization $\mathfrak{p} = \mathfrak{z} + \mathfrak{u}$. That demonstrates Theorem 3.1(1).

If $f(\mathfrak{z}) = 0$ then $\pi_{f,\mathfrak{p}}$ is a unitary character on H_n, automatically irreducible. Now suppose $f(\mathfrak{z}) \neq 0$ and $\mathfrak{p} = \mathfrak{z} + \mathfrak{u}$. Then $\pi_{f,\mathfrak{p}}$ is a representation of H_n on $L^2(G/P) = L^2(W)$ where $W = \exp(\mathfrak{w}) \cong \mathbb{R}^n$. Then $\pi_{f,\mathfrak{p}}(H_n)$ acts by all translations on W and by scaling that distinguishes the integrands of $L^2(W) = \int_{\mathfrak{w}^*} e^{\langle \xi, \cdot\rangle}\mathbb{C}\,d\xi$, so it is irreducible. That demonstrates Theorem 3.1(2).

If $f(\mathfrak{z}) = 0$ then $\mathfrak{h}_n$ is the only real polarization for f. Now suppose $f(\mathfrak{z}) \neq 0$ and consider the case where $\mathfrak{p} = \mathfrak{z} + \mathfrak{u}$ and $\mathfrak{p}' = \mathfrak{z} + \mathfrak{w}$. Then $\omega = \mathrm{Im}\,h(\cdot, \cdot)$ pairs $\mathfrak{u}$ with $\mathfrak{w}$, and the Fourier transform $\mathcal{F}$: $L^2(U) \cong L^2(W)$ intertwines $\pi_{f,\mathfrak{p}}$ with $\pi_{f',\mathfrak{p}}$. More generally, if $\mathfrak{p}$ and $\mathfrak{p}'$ are any two real polarizations for f, then we write $\mathfrak{p} = (\mathfrak{p} \cap \mathfrak{p}') + \mathfrak{u}'$, $\mathfrak{p}' = (\mathfrak{p} \cap \mathfrak{p}') + \mathfrak{w}'$ and $\mathfrak{p} \cap \mathfrak{p}' = \mathfrak{z} + \mathfrak{v}'$. That done, ω pairs $\mathfrak{u}'$ with $\mathfrak{w}'$, and the corresponding Fourier transform $\mathcal{F}$: $L^2(U') \cong L^2(W')$ combines with the identity transformation of $L^2(V')$ to give a map $L^2(V')\widehat{\otimes}L^2(U') \cong L^2(V')\widehat{\otimes}L^2(W')$ that intertwines $\pi_{f,\mathfrak{p}}$ with $\pi_{f',\mathfrak{p}}$. This demonstrates Theorem 3.1(3).

Now Theorem 3.1(4) follows from the considerations we outlined in Section 2. In the terminology there, the infinite dimensional irreducible unitary representation π_ζ of H_n is equivalent to π_f whenever $f \in \mathfrak{n}^*$ such that $f(z) = \langle \zeta, x\rangle$ for every $z \in \mathfrak{z}$. In particular, if $f(\mathfrak{z}) \neq 0$ where $\mathfrak{z}$ is the center of the Heisenberg algebra, then the coadjoint orbit $\mathcal{O}_f = f + \mathfrak{z}^\perp$, where $\mathfrak{z}^\perp := \{h \in \mathfrak{n}^* \mid h(\mathfrak{z}) = 0\}$. Of course one can also verify this by direct computation.

4. Square integrable representations

In this section N is a connected simply connected nilpotent Lie group and Z is its center. If $\zeta \in \widehat{Z}$ we denote $\widehat{N}_\zeta = \{[\pi] \in \widehat{N} \mid \pi|_Z \text{ is a multiple of } \zeta\}$.

The corresponding L^2 space is

$$L^2(N/Z : \zeta)$$
$$:= \left\{ f : N \to \mathbb{C} \text{ measurable} \ \middle| \ \begin{array}{l} f(nz) = \zeta(z)^{-1} f(n) \text{ and} \\ \int_{N/Z} |f(n)|^2 d\mu_{N/Z}(nZ) < \infty \end{array} \right\}. \quad (4.1)$$

The inner product $\langle f, h \rangle_\zeta = \int_{N/Z} f(n)\overline{h(n)} d\mu_{N/Z}(nZ)$ is well defined on the relative L^2 space $L^2(N/Z : \zeta)$. Each $\widehat{N}_\zeta$ is a measurable subset of $\widehat{N}$, and $\widehat{N} = \bigcup_{\zeta \in \widehat{Z}} \widehat{N}_\zeta$. Here $L^2(N) = \int_{\widehat{Z}} L^2(N/Z : \zeta)$. This decomposes the left regular representation of N as

$$\ell = \operatorname{Ind}_{\{1\}}^N (1) = \operatorname{Ind}_Z^N \operatorname{Ind}_{\{1\}}^Z (1) = \operatorname{Ind}_Z^N \int_{\widehat{Z}} \zeta \, d\zeta$$

$$= \int_{\widehat{Z}} \operatorname{Ind}_Z^N \zeta \, d\zeta = \int_{\widehat{Z}} \ell_\zeta \, d\zeta$$

where $\ell_\zeta = \operatorname{Ind}_Z^N \zeta$ is the left regular representation of N on $L^2(N/Z : \zeta)$. The corresponding expansion for functions,

$$f(n) = \int_{\widehat{Z}} f_\zeta(n) d\zeta \text{ where } f_\zeta(n) = \int_Z f(nz)\zeta(z)d\mu_Z(z),$$

is just Fourier inversion on the commutative locally compact group Z.

Now we describe some results of Moore and myself on square integrable representations in this context. The first observation is

Theorem 4.2. *Let N be a connected simply connected nilpotent Lie group and $\zeta \in \widehat{Z}$. If $[\pi] \in \widehat{N}_\zeta$ then the following conditions are equivalent.*

1. *There exist nonzero $u, v \in H_\pi$ such that $|f_{u,v}| \in L^2(N/Z)$, i.e., $f_{u,v} \in L^2(N/Z : \zeta)$.*
2. *The coefficient $|f_{u,v}| \in L^2(N/Z)$, equivalently $f_{u,v} \in L^2(N/Z : \zeta)$, for all $u, v \in H_\pi$.*
3. *$[\pi]$ is a discrete summand of ℓ_ζ.*

A representation class $[\pi] \in \widehat{N}$ is L^2 or **square integrable** or **relative discrete series** if its coefficients $f_{u,v}(n) = f_{\pi:u,v}(n) := \langle u, \pi(n)v \rangle$ satisfy $|f_{u,v}| \in L^2(N/Z)$, in other words if its coefficients are square integrable modulo Z. Theorem 4.2 says that it is sufficient to check this for just one nonzero coefficient, and Theorem 4.2(3) justifies the term "relative discrete series".

We say that N **has square integrable representations** if at least one class $[\pi] \in \widehat{N}$ is square integrable. These representations satisfy an analog of the Schur orthogonality relations:

Theorem 4.3. *Let N be a connected simply connected nilpotent Lie group. If $\zeta \in \widehat{Z}$ and $[\pi] \in \widehat{N}_\zeta$ is square integrable then there is a number $\deg(\pi) > 0$ such that the coefficients of π satisfy*

$$\int_{N/Z} f_{u_1,v_1}(n)\overline{f_{u_2,v_2}(n)}d\mu_{N/Z}(nZ) = \frac{1}{\deg(\pi)}\langle u_1, u_2\rangle\overline{\langle v_1, v_2\rangle} \qquad (4.4)$$

for all $u_i, v_i \in H_\pi$. If $[\pi_1], [\pi_2] \in \widehat{N}_\zeta$ are inequivalent square integrable representations then their coefficients are orthogonal in $L^2(N/Z : \zeta)$,

$$\int_{N/Z} f_{\pi_1:u_1,v_1}(n)\overline{f_{\pi_2:u_2,v_2}(n)}d\mu_{N/Z}(nZ) = 0, \qquad (4.5)$$

for all $u_1, v_1 \in H_{\pi_1}$ and $u_2, v_2 \in H_{\pi_2}$.

The number $\deg(\pi)$ is the **formal degree** of $[\pi]$. It plays the same role in Theorem 4.3 as that played by the degree in the Schur orthogonality relations for compact groups . In general, $\deg(\pi)$ depends on normalization of Haar measure: a rescaling of Haar measure $\mu_{N/Z}$ of N/Z to $c\mu_{N/Z}$ rescales formal degrees $\deg(\pi)$ to $\frac{1}{c}\deg(\pi)$. We don't see this for compact groups because there we always scale Haar measure to total mass 1.

Theorems 4.2 and 4.3 only require that N be a locally compact group of Type I and that Z be a closed subgroup of the center of N. They can be understood as special cases of Hilbert algebra theory. Here we related them to the Kirillov theory.

Given $f \in \mathfrak{n}^*$ we have the bilinear form $b_f(x, y) = f([x, y])$, the coadjoint orbit $\mathcal{O}_f = \mathrm{Ad}^*(N)f$, the associated representation $[\pi_f]$, and the character $\zeta \in \widehat{Z}$ such that $[\pi_f] \in \widehat{N}_\zeta$. Note that $f|_{\mathfrak{z}}$ determines the affine subspace $f + \mathfrak{z}^\perp$ in $\mathfrak{n}^*$.

Theorem 4.6. *Let N be a connected simply connected nilpotent Lie group and $f \in \mathfrak{n}^*$. Then the following conditions are equivalent.*
1. *$[\pi_f]$ is square integrable.*
2. *The left regular representation ℓ_ζ of N on $L^2(N/Z : \zeta)$ is primary.*
3. *$\mathcal{O}_f = f + \mathfrak{z}^\perp$, determined by the restriction $f|_{\mathfrak{z}}$.*
4. *b_f is nondegenerate on $\mathfrak{n}/\mathfrak{z}$.*

Recall the notion of the Pfaffian $\mathrm{Pf}(\omega)$ of an antisymmetric bilinear form ω on a finite dimensional real vector space V relative to a volume form ν on V. If $\dim V$ is odd then by definition $\mathrm{Pf}(\omega) = 0$. If $\dim V = 2m$ even, then ω^m is a multiple of ν, and by definition that multiple of $\mathrm{Pf}(\omega)$; in other words $\omega^m = \mathrm{Pf}(\omega)\nu$. The Pfaffian is the square root of the determinant on antisymmetric bilinear forms.

Fix a volume element ν on $\mathfrak{v} := \mathfrak{n}/\mathfrak{z}$. If $f \in \mathfrak{n}^*$ we view $\omega_f(x, y) = f([x, y])$ as an antisymmetric bilinear form on $\mathfrak{v}$. Define $P(f) := \text{Pf}(\omega_f)$. Then P is a homogeneous polynomial function on $\mathfrak{n}^*$, and $P(f)$ depends only on $f|_{\mathfrak{z}}$ a So there is a homogeneous polynomial function (which we also denote P) on $\mathfrak{z}^*$ such that $P(f) = P(f|_{\mathfrak{z}})$.

In the case of the Heisenberg group H_m, P is the homogeneous polynomial $P(\zeta) = \zeta(z_0)^m$ of degree m on $\mathfrak{z}^*$. Here the choice of nonzero $z_0 \in \mathfrak{z}$ is a normalization, in effect a choice of unit vector. As described in Theorem 4.9 below, this also gives the formal degree of $[\pi_f]$ where $\zeta = f|_{\mathfrak{z}}$. Further, as described in Theorem 4.11, it gives the Plancherel measure on $\widehat{H_m}$.

In view of Theorem 4.6 we now have

Theorem 4.7. *The representation π_f is square integrable if and only if the Pfaffian polynomial $P(f|_{\mathfrak{z}}) \neq 0$. In particular $\phi : f|_{\widehat{\mathfrak{z}}} \mapsto [\pi_f]$ defines a bijection from $\{\lambda \in \mathfrak{z}^* \mid P(\lambda) \neq 0\}$ onto $\{[\pi] \in \widehat{N} \mid [\pi] \text{ is square integrable}\}$.*

One can view the polynomial P as an element of the symmetric algebra $S(\mathfrak{z})$, and since $\mathfrak{z}$ is commutative that symmetric algebra is the same as the universal enveloping algebra $\mathfrak{Z}$. From this one can prove

Corollary 4.8. *The group N has square integrable representations if and only if the inclusion $\mathfrak{Z} \hookrightarrow \mathfrak{N}$ of universal enveloping algebras, induced by $\mathfrak{z} \hookrightarrow \mathfrak{n}$, maps $\mathfrak{Z}$ onto the center of $\mathfrak{N}$.*

Both formal degree and the polynomial P are scaled by $1/c$ when Haar measure on N/Z is scaled by c, so the following is independent of normalization of Haar measure on N/Z.

Theorem 4.9. *The formal degree of a square integrable representation $[\pi_f] = \phi(f|_{\mathfrak{z}})$ is given by $\deg(\pi_f) = |P(f|_{\mathfrak{z}})|$.*

As in the semisimple case, the **infinitesimal character** of a representation class $[\pi] \in \widehat{N}$ is the associative algebra homomorphism $\xi_\pi : Cent(\mathfrak{N}) \to \mathbb{C}$ from the center of the enveloping algebra, such that $d\pi(\zeta)$ is scalar multiplication by $\chi_\pi(\zeta)$. (Initially this holds only on C^∞ vectors, but they are dense in H_π, so by continuity it holds on all vectors.) If $\zeta \in \mathfrak{z}$ then $\chi_{\pi_f}(\zeta) = if(\zeta)$. Now, from Theorem 4.9,

Corollary 4.10. *If $[\pi] \in \widehat{N}$ then the formal degree $\deg(\pi) = |\chi_\pi(P)|$ where we understand the formal degree of a non square integrable representation to be zero.*

For the Plancherel formula and Fourier inversion we must normalize Haar measures. Choose Haar measures μ_Z and $\mu_{N/Z}$; they define a Haar measure μ_N by $d\mu_N = d\mu_{N/Z}\,d\mu_Z$, i.e.

$$\int_N f(n)d\mu_N(n) = \int_{N/Z} \left(\int_Z f(nz)d\mu_Z(z) \right) d\mu_{N/Z}(nZ).$$

Now we have Lebesgue measures ν_Z, $\nu_{N/Z}$ and ν_N on $\mathfrak{z}$, $\mathfrak{n}/\mathfrak{z}$ and $\mathfrak{n}$ speci-fied by the condition that the exponential map have Jacobian 1 at 0, and they satisfy $d\nu_N = d\nu_{N/Z}\,d\nu_Z$. Normalize Lebesgue measures on the dual spaces by the condition that Fourier transform is an isometry; that gives Lebesgue measures ν_Z^*, $\nu_{N/Z}^*$ and ν_N^* such that $d\nu_N^* = d\nu_{N/Z}^*\,d\nu_Z^*$.

Theorem 4.11. *Let N have square integrable representations. Let $c = m!2^m$ where $2m$ is the maximum dimension of the $\mathrm{Ad}^*(N)$–orbits in $\mathfrak{n}^*$. Then Plancherel measure for N is concentrated on the square integrable classes and its image under the map*

$$\phi^{-1} : \{[\pi] \in \widehat{N} \mid [\pi] \text{ is square integrable}\} \to \{\lambda \in \mathfrak{z}^* \mid P(\lambda) \neq 0\}$$

of Theorem 4.7 *is $c|P(x)|d\nu_Z^*(x)$.*

5. Commutative spaces – generalities

We just described the Fourier transform and Fourier inversion formulae for H_m — and a somewhat larger class of connected simply connected nilpotent Lie groups. Now we edge toward a more geometric setting, that of commutative spaces, which is a common generalization of Rie-mannian symmetric spaces, locally compact abelian groups and homo-geneous graphs. It is interesting and precise for the cases that involve connected simply connected nilpotent Lie groups with square integrable representations.

A *commutative space* G/K, equivalently a *Gelfand pair* (G, K), con-sists of a locally compact group G and a compact subgroup K such that the convolution algebra $L^1(K\backslash G/K)$ is commutative. There are several other formulations. Specifically, the following are equivalent.

1. (G, K) is a Gelfand pair, i.e. $L^1(K\backslash G/K)$ is commutative under con-volution.
2. If $g, g' \in G$ then $\mu_{KgK} * \mu_{Kg'K} = \mu_{Kg'K} * \mu_{KgK}$ (convolution of mea-sures on $K\backslash G/K$).
3. $C_c(K\backslash G/K)$ is commutative under convolution.
4. The measure algebra $\mathcal{M}(K\backslash G/K)$ is commutative.
5. The (left regular) representation of G on $L^2(G/K)$ is multiplicity free.

If G is a connected Lie group one can add

6. The algebra $\mathcal{D}(G, K)$ of G–invariant differential operators on G/K is commutative.

Commutative spaces G/K are important for a number of reasons. First, they are manageable because their basic harmonic analysis is very similar to that of locally compact abelian groups. We will describe that in a moment. Second, in the Lie group cases, most of the G/K carry invariant weakly symmetric Riemannian metrics, and have properties very similar to those of Riemannian symmetric spaces. Third, the invariant differential operators and corresponding spherical functions play a definite role in special function theory. And fourth, in the nilpotent Lie groups setting, there is some interesting interplay between geometry and hypoellipticity.

We only consider the basic harmonic analysis, here in general and in Section 6 for the case of commutative nilmanifolds.

Analysis on locally compact abelian groups is based on decomposition of functions in terms of unitary characters. In the classical euclidean case these are just the complex exponentials $\chi_\xi : V \to \mathbb{C}, \xi \in V^*$, given by $\chi_\xi = e^{ix \cdot \xi}$. For a commutative space G/K the appropriate replacements are the positive definite spherical functions, defined as follows.

A continuous K–bi–invariant function $\varphi : G \to \mathbb{C}$ is K–$spherical$ if $\varphi(1) = 1$ and $f \mapsto (f * \varphi)(1)$ is a homomorphism $C_c(K \backslash G/K) \to \mathbb{C}$. Equivalent: φ is not identically zero, and if $g_1, g_2 \in G$ then

$$\varphi(g_1)\varphi(g_2) = \int_K \varphi(g_1 k g_2)dk.$$

A function $\varphi : G \to \mathbb{C}$ is $positive\ definite$ if $\sum \varphi(g_j^{-1}g_i)\overline{c_j}c_i \geqq 0$ whenever $\{c_1, \ldots, c_n\} \subset \mathbb{C}$ and $\{g_1, \ldots, g_n\} \subset G$.

Denote $\mathcal{P} = \mathcal{P}(G, K)$: positive definite K–spherical functions on G. There is a one–one correspondence $\varphi \leftrightarrow \pi_\varphi$ between $\mathcal{P}$ and the irreducible unitary representations π of G that have a K–fixed unit vector v. It is given by $\varphi(g) = \langle v, \pi(g)v \rangle_{\mathcal{H}_\pi}$. We have the $spherical\ transform$

$$\mathcal{S} : f \mapsto \widehat{f} \text{ from } L^1(K \backslash G/K) \text{ to functions on } \mathcal{P}$$

defined by

$$\mathcal{S}(f)(\varphi) = \widehat{f}(\varphi) = (f * \varphi)(1) = \int_G f(g)\varphi(g^{-1})dg.$$

The corresponding spherical inversion formula is

$$f(g) = \int_{\mathcal{P}} \widehat{f}(\varphi)\varphi(g)d\mu(\varphi).$$

Here $\mathcal{P}$ has natural structure of locally compact space and μ is called Plancherel measure. The spherical transform

$$\mathcal{S} : L^1(K\backslash G/K) \cap L^2(K\backslash G/K) \to L^2(\mathcal{P}, \mu)$$

preserves L^2 norm and extends by continuity to an isometry

$$\mathcal{S} : L^2(K\backslash G/K) \cong L^2(\mathcal{P}, \mu).$$

Note that $\mathcal{S}$ can only be given by its defining integral expression when that integral converges. This is why it has to be extended by L^2 continuity. Of course this problem is already present with the classical Fourier transform on $\mathbb{R}$.

The Plancherel Formula $\mathcal{S} : L^2(K\backslash G/K) \cong L^2(\mathcal{P}, \mu)$ gives a continuous direct sum (direct integral) decomposition

$$L^2(K\backslash G/K) \cong \int_{\mathcal{P}} \mathbb{C}\varphi \, d\mu(\varphi).$$

This extends to a continuous direct sum decomposition

$$L^2(G/K) \cong \int_{\mathcal{P}} \mathcal{H}_{\pi_\varphi} \, d\mu(\varphi).$$

Of course all this depends on knowledge of the Plancherel measure μ.

6. Commutative nilmanifolds

Theorem of Carcano (special case): Let $K \subset U(m)$ acting on $\mathbb{C}^m$, where $\mathfrak{h}_m = \mathrm{Im}\,\mathbb{C} + \mathbb{C}^m$ with center $\mathrm{Im}\,\mathbb{C}$. Then $(H_m \rtimes K)/K$ is commutative if and only if the representation of $K_{\mathbb{C}}$, on the ring of all polynomials on $\mathbb{C}^m$, is multiplicity free.

Kač classified the connected K_m that are irreducible on $\mathbb{C}^m$:

	Group K_m	Acting on		Group K_m	Acting on
1	$SU(m)$	$\mathbb{C}^m, m \geqq 2$	8	$U(n)$	$\mathbb{C}^m = \Lambda^2(\mathbb{C}^n)$
2	$U(m)$	$\mathbb{C}^m, m \geqq 1$	9	$SU(\ell) \times SU(n)$	$\mathbb{C}^m = \mathbb{C}^\ell \otimes \mathbb{C}^n, \ell \neq n$
3	$Sp(n)$	$\mathbb{C}^m = \mathbb{C}^{2n}$	10	$S(U(\ell) \times U(n))$	$\mathbb{C}^m = \mathbb{C}^\ell \otimes \mathbb{C}^n$
4	$U(1) \times Sp(n)$	$\mathbb{C}^m = \mathbb{C}^{2n}$	11	$U(2) \times Sp(n)$	$\mathbb{C}^m = \mathbb{C}^2 \otimes \mathbb{C}^{2n}$
5a	$U(1) \times SO(2n)$	$\mathbb{C}^m = \mathbb{C}^{2n}$	12	$SU(3) \times Sp(n)$	$\mathbb{C}^m = \mathbb{C}^3 \otimes \mathbb{C}^{2n}$
5b	$U(1) \times SO(2n+1)$	$\mathbb{C}^m = \mathbb{C}^{2n+1}$	13	$U(3) \times Sp(n)$	$\mathbb{C}^m = \mathbb{C}^3 \otimes \mathbb{C}^{2n}$
6	$U(n), n \geqq 2$	$\mathbb{C}^m = S^2(\mathbb{C}^n)$	14	$SU(n) \times Sp(4)$	$\mathbb{C}^m = \mathbb{C}^n \otimes \mathbb{C}^8$
7	$SU(n), n$ odd	$\mathbb{C}^m = \Lambda^2(\mathbb{C}^n)$	15	$U(n) \times Sp(4)$	$\mathbb{C}^m = \mathbb{C}^n \otimes \mathbb{C}^8$

A *commutative nilmanifold* is a commutative space G/K where some connected closed nilpotent subgroup of G is transitive on G/K.

Example: G/K where $G = H_m \rtimes K_m$ and $K = K_m$, where K_m occurs in the table above.

Fact: Let G/K be commutative. If a conn closed nilpotent subgroup N of G is transitive then N is the nilradical of G, N is abelian or 2–step nilpotent, and $G = N \rtimes K$.

In particular: Commutative nilmanifolds have form G/K where $G/K = (N \rtimes K)/K$, N is not so different from the Heisenberg group and $K \subset \text{Aut}(N)$.

More examples: a commutative nilmanifold $(G = N \rtimes K)/K$ is *irreducible* if K acts irreducibly on $\mathfrak{n}/[\mathfrak{n}, \mathfrak{n}]$, *maximal* if it is not of the form $(\widetilde{G}/\widetilde{Z}, \widetilde{K}/\widetilde{Z})$ with $\{1\} \neq \widetilde{Z} \subset \widetilde{K}$ central in $\widetilde{G}$. They have been classified by Vinberg. Let $\mathfrak{n} = \mathfrak{z} + \mathfrak{w}$ where $\mathfrak{z}$ is the center and $\text{Ad}(K)\mathfrak{w} = \mathfrak{w}$. Then $\mathfrak{z}$ is the center of $\mathfrak{n}$, $\mathfrak{w} \cong \mathfrak{n}/[\mathfrak{n}, \mathfrak{n}]$ as K–module, and the classification is

	Group K	$\mathfrak{w}$	$\mathfrak{z}$
1	$SO(n)$	$\mathbb{R}^n$	Skew $\mathbb{R}^{n \times n} = \mathfrak{so}(n)$
2	$Spin(7)$	$\mathbb{R}^8 = \mathbb{O}$	$\mathbb{R}^7 = \text{Im}\,\mathbb{O}$
3	G_2	$\mathbb{R}^7 = \text{Im}\,\mathbb{O}$	$\mathbb{R}^7 = \text{Im}\,\mathbb{O}$
4	$U(1) \cdot SO(n), n \neq 4$	$\mathbb{C}^n$	$\text{Im}\,\mathbb{C}$
5	$(U(1)\cdot)SU(n)$	$\mathbb{C}^n$	$\Lambda^2 \mathbb{C}^n \oplus \text{Im}\,\mathbb{C}$
6	$SU(n), n$ odd	$\mathbb{C}^n$	$\Lambda^2 \mathbb{C}^n$
7	$SU(n), n$ odd	$\mathbb{C}^n$	$\text{Im}\,\mathbb{C}$
8	$U(n)$	$\mathbb{C}^n$	$\text{Im}\,\mathbb{C}^{n \times n} = \mathfrak{u}(n)$
9	$(U(1)\cdot)Sp(n)$	$\mathbb{H}^n$	$\text{Re}\,\mathbb{H}_0^{n \times n} \oplus \text{Im}\,\mathbb{H}$
10	$U(n)$	$S^2\mathbb{C}^n$	$\mathbb{R}$
11	$(U(1)\cdot)SU(n), n \geq 3$	$\Lambda^2\mathbb{C}^n$	$\mathbb{R}$
12	$U(1) \cdot Spin(7)$	$\mathbb{C}^8$	$\mathbb{R}^7 \oplus \mathbb{R}$
13	$U(1) \cdot Spin(9)$	$\mathbb{C}^{16}$	$\mathbb{R}$
14	$(U(1)\cdot)Spin(10)$	$\mathbb{C}^{16}$	$\mathbb{R}$
15	$U(1) \cdot G_2$	$\mathbb{C}^7$	$\mathbb{R}$
16	$U(1) \cdot E_6$	$\mathbb{C}^{27}$	$\mathbb{R}$
17	$Sp(1) \times Sp(n), n \geq 2$	$\mathbb{H}^n$	$\text{Im}\,\mathbb{H} = \mathfrak{sp}(1)$
18	$Sp(2) \times Sp(n)$	$\mathbb{H}^{2 \times n}$	$\text{Im}\,\mathbb{H}^{2 \times 2} = \mathfrak{sp}(2)$
19	$(U(1)\cdot)SU(m) \times SU(n)$ $m, n \geq 3$	$\mathbb{C}^m \otimes \mathbb{C}^n$	$\mathbb{R}$
20	$(U(1)\cdot)SU(2) \times SU(n)$	$\mathbb{C}^2 \otimes \mathbb{C}^n$	$\text{Im}\,\mathbb{C}^{2 \times 2} = \mathfrak{u}(2)$
21	$(U(1)\cdot)Sp(2) \times SU(n), n \geq 3$	$\mathbb{H}^2 \otimes \mathbb{C}^n$	$\mathbb{R}$
22	$U(2) \times Sp(n)$	$\mathbb{C}^2 \otimes \mathbb{H}^n$	$\text{Im}\,\mathbb{C}^{2 \times 2} = \mathfrak{u}(2)$
23	$U(3) \times Sp(n), n \geq 2$	$\mathbb{C}^3 \otimes \mathbb{H}^n$	$\mathbb{R}$

where the optional $U(1)$ is required in (5) when n is odd, in (11) when n is even, in (19) when $m = n$, in (20) when $n = 2$, and in (21) when $n \leq 4$. Here (9) was the first known case where G/K is not weakly symmetric (Lauret).

To make this explicit one needs to know the positive definite spherical functions and the Plancherel measure.

In the connected Lie group cases, the K–spherical functions on G are just the joint eigenfunctions for $\mathcal{D}(G, K)$, and in many cases this is how one finds them. We look at a few of those cases.

Case $(G, K) = (\mathbb{R}^n \rtimes K, K)$ where K is transitive on the unit sphere in $\mathbb{R}^n$. Then the invariant differential operators are the polynomials in the Laplacian $\Delta = -\sum \partial^2/\partial x_i^2$, and the K–spherical functions are the radial eigenfunctions of Δ for real non-negative eigenvalue. They are the

$$\varphi_\xi(x) = (\|\xi\| r)^{-(n-2)/2} J_{(n-2)/2}(\|\xi\| r)$$

where $r = \|x\|$ and J_ν is the Bessel function of first kind and order ν.

Case $(G, K) = (H_n \rtimes U(n), U(n))$. In the coordinate $(z, w) \in \operatorname{Im}\mathbb{C} + \mathbb{C}^n = H_n$ the invariant differential operators are the polynomials in $\partial/\partial z$, $\square = -\sum \partial^2/\partial w_i^2$ and $\overline{\square}$. The positive definite spherical functions corresponding to 1–dimensional representations are the

$$\varphi_\xi(z, w, k) = \frac{2^{n-1}(n-1)!}{(\|\xi\|\,\|w\|)^{n-1}} J_{n-1}(\|\xi\|\,\|w\|) \text{ for } 0 \neq \xi \in \mathfrak{w}^*$$

and those for infinite dimensional representations are the

$$\varphi_{\zeta,m} : (z, w, k) \mapsto \begin{cases} e^{i\zeta(z)} L_m^{(n-1)}(\zeta(z)\|w\|^2) e^{-\zeta(z)\|v\|^2/4} & \text{if } \zeta(i) > 0, \\ \varphi_{-\zeta,m}(z, w, k) & \text{if } \zeta(i) < 0, \end{cases}$$

where $\zeta \in (\operatorname{Im}\mathbb{C})^*$ and $L_m^{(n-1)}$ is the generalized Laguerre polynomial of order $n - 1$ normalized to $L_m^{(n-1)}(0) = 1$.

References

[1] J. A. WOLF, "Harmonic Analysis on Commutative Spaces", Mathematical Surveys & Monographs **142**, American Math. Soc., 2007.

[2] J. A. WOLF, *Infinite dimensional multiplicity–free spaces III: Matrix coefficients and regular functions*, Mathematische Annalen **349** (2011), 263–299.

Leibniz' conjecture, periods & motives

Gisbert Wüstholz

Questions on the transcendence and linear independence of periods have a long history going back at least to Euler. We shall first give in this note a historical introduction to periods with the aim to demonstrate how a very nice and deep theory evolved during 3 centuries with three themes: numbers (Euler, Leibniz, Hermite, Lindemann, Siegel, Gelfond, Schneider, Baker), Hodge theory (Hodge, De Rham, Grothendieck, Griffiths, Deligne) and motives (Deligne, Nori). One of our main intends is to discuss then how to possibly bring these themes together and to show how modern transcendence theory can solve questions which arise at the interfaces between number theory, global analysis, algebraic geometry and arithmetic algebraic geometry.

1. Basic discourse

Let X be a smooth projective variety defined over a number field $K \subset \mathbb{C}$ and let D be an ample divisor on X with normal crossings. We denote by U the complement of D and by Ω_U^1 the sheaf of holomorphic 1-forms on U. Let ξ be a closed differential form in $\Gamma(U, \Omega_U^1)$ and $\gamma : I \to U_{\mathbb{C}} = U \times_K \mathbb{C}$ a path. We consider the integral $\int_\gamma \xi$ and we shall study transcendence properties of the integral in the case when $\gamma(i) \in U(K)$ for $i = 0, 1$. The problem has an old history dating back to the work of Euler and of Leibniz and surprisingly it turns out that the integral can take transcendental but also algebraic values. The question then is: when is the value transcendental and when is it algebraic?

The answer depends, as it turns out, on the differential form and on the path as can be easily seen from examples: the differential form could be exact. Then obviously the integral takes algebraic values. Non-trivial examples can be obtained in the following way. Let E be an elliptic curve defined over a number field and let $\Gamma \subset E \times E$ be the graph of an endomorphism φ of E. Then the differential form $\xi = pr_1^* \varphi^* dx/y - pr_2^* dx/y$ vanishes on Γ so that the integral $I(\xi, \gamma)$ is zero when the path γ is contained in Γ. If we modify ξ by some exact form df and make sure that

γ does not contain any pole of f then $I(\xi, \gamma)$ takes algebraic values, possibly non-zero.

A more interesting example is obtained by taking the elliptic curve E in $\mathbb{P}^2$ with equation $T^2 = (S^2 - 1)(S - t)$ where $t \neq 0, \pm 1, \infty$ is in $\mathbb{P}^1$. The curve X in $\mathbb{P}^2$ with equation $T^2 = (S^4 - 1)(S^2 - t)$ has genus 2 and there is a morphism $\alpha : X \to E$ given by $(S, T) \mapsto (S^2, T)$. The morphism α induces a homomorphism $J(\alpha) : J(X) \to E$ between the Jacobians. We denote by $\kappa : E' \to J(X)$ the kernel of $J(\alpha)$. The holomorphic 1-form dx/y on the elliptic curve E pulls back to a holomorphic 1-form $\omega = J(\alpha)^* dx/y$ on $J(X)$. Let $\iota : X \to J(X)$ be the canonical embedding of X into the Jacobian. This leads to a commutative diagram

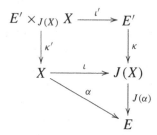

with $E' \times_{J(X)} X$ a closed subscheme of X and of E'.

We choose a path γ with class $[\gamma]$ in the relative homology $H_1(X, E' \times_{J(X)} X, \mathbb{Z})$ of X with respect to $E' \times_{J(X)} X$ such that $\gamma(0)$ is the base point of X. If $\iota_*[\gamma]$ is in the image under κ of the relative homology $H_1(E', E' \times_{J(X)} X, \mathbb{Z})$ of E' with respect to $E' \times_{J(X)} X$ then $\iota_*[\gamma]$ can be written as $\kappa_*[\delta]$ for some $[\delta] \in H_1(E', E' \times_{J(X)} X, \mathbb{Z})$ and we get

$$\int_\gamma \alpha^*(dx/y) = \int_\gamma \iota^* \circ J(\alpha)^*(dx/y) = \int_{\iota_*\gamma} J(\alpha)^*(dx/y) = \int_{\kappa_*\delta} J(\alpha)^*(dx/y)$$

$$= \int_\delta \kappa^* \circ J(\alpha)^*(dx/y) = \int_\delta (J(\alpha) \circ \kappa)^*(dx/y) = 0.$$

The 1-form $\xi = \iota^*\omega \in H^0(X, \Omega_X^1)$ is closed and holomorphic and defines a multi-valued function $I(\xi, \gamma) = \int_\gamma \xi$ in γ with zeroes exactly when $\iota_*[\gamma] = \kappa_*[\delta]$ for $[\delta] \in H_1(E', E' \times_{J(X)} X, \mathbb{Z})$ which is a finite set.

The same modification by an exact form as above shows that $I(\xi, \gamma)$ can take also non-zero algebraic values. However it can be shown that for fixed ξ zeroes γ of the multivalued transcendental function I_ξ which takes $\gamma : [0, 1] \to X(\mathbb{C})$ into $I(\xi, \gamma)$ have the property that, up to finitely many effectively computable exceptions, the boundary $\partial\gamma$ is not in $\overline{\mathbb{Q}}$ or equivalently that at least one out of $\delta(0)$ and $\delta(1)$ is transcendental.

We see that the question whether $I(\xi, \gamma)$ is algebraic or not is a very subtle problem and we are looking for a complete description of the exceptional set of paths γ in U for which $I(\xi, \gamma)$ is algebraic. This set is a subset of the so-called mapping space $\mathcal{P}_U$, an infinite dimensional manifold, and we shall see that the set can be determined in an intrinsic topological way.

2. Historical background

In this note we study transcendence properties of rational integrals on projective varieties. Our investigations were motivated by the fact that many numbers which were proved in the past to be transcendental can be written as rational intergrals with algebraic bounds. We shall discuss a general result formulated here as a conjecture which contains the results we were alluding to as special cases. Another motivation was coming from the very interesting monograph *Huygens and Barrow, Newton and Hook* of Arno'ld where we found a reference to a letter of Leibniz to Huygens, dated $\frac{10}{20}$ April 1691. In this letter Leibniz formulated the problem of transcendence of the areas of segments cut off from an algebraic curve, defined by an equation with rational coefficients, by straight lines with algebraic coefficients (see [1], page 93, footnote). In his book Arnold reformulated this problem into modern language as *an abelian integral along an algebraic curve with rational (algebraic) coefficients taken between limits which are rational (algebraic) numbers is generally a transcendental number* and it is the word *generally* which initiated our research and our intension was to clarify the situation in a conceptual way.

The most famous period is $2\pi i$ which is related to the old Greek problem on squaring the circle by ruler and compass and which was open for more than 2000 years. The impossibility of doing so was then proved in 1882 by F. Lindemann in his famous article where he shows that π is transcendental. The result can be formulated also in this way: non-zero periods of the differential form dx/x are transcendental. Actually he proves more. Namely he shows that if $\alpha \neq 0$ not both of the numbers α and $\exp \alpha$ can be algebraic. The transcendenc of π is deduced by putting $\alpha = i\pi$. Again this corresponds to a result on transcendence properties of rational integrals. Namely we take the differential dx/x and integrate from 1 to $\exp \alpha$. If $\exp \alpha$ is algebraic then

$$\alpha = \int_1^{\exp \alpha} \frac{dx}{x}$$

must be transcendental according to Lindemann's result.

Historically the next significant progress in the general problem was made by Carl Ludwig Siegel who showed that non-zero periods of elliptic integrals of the first kind with complex multiplication are transcendental. The result was superseded by the result of Th. Schneider who proved that elliptic integrals of the second kind taken between algebraic bounds are either zero or transcendental without the hypothesis of complex multiplication. Shortly after he also suceeded to extend Siegel's result to abelian integrals of the second kind. As a nice example on can deduce the transcendence of values of the Beta-function

$$B(a, b) = \frac{\Gamma(a)\Gamma(b)}{\Gamma(a + b)}$$

at rational arguments a, b, such that none of the numbers $a, b, a + b$ are integral. In fact these Beta-values are periods of differentials of the second kind on curves whose Jacobians are abelian varieties of CM-type and so Schneider's results apply.

In the abelian case the first result without the hypothesis of complex multiplication in the very special case of a product of two elliptic curves was obtained by A. Baker and extended by J. Coates and D. Masser later to cover, still in this particular situation, integrals of the second kind completely. The more general case dealing with a product of an arbitrary number of elliptic curves was solved by the author and this solved a problem of A. Baker.

After Lindemann's discovery Hilbert asked in his famous address to the international congress of mathematicians at Paris in 1900 the question whether numbers of the form α^{β} for algebraic numbers α and β are transcendental except that α is zero or one or β is rational. This problem was solved independently by A. O. Gel'fond and Th. Schneider. The result can be transformed into a question on rational integrals so to fit into our general point of view: consider the affine plane X with coordinates x and y, the differential form $\omega = (dx/x) - \beta(dy/y)$ and a path $\gamma : [0, 1] \rightarrow X$ with $\gamma(0) = (1, 1)$ and $\gamma(1) = (\delta, \alpha)$. Then the integral of the differential form ω along the path is $\log \delta - \beta \log \alpha$ and the theorem of Gel'fond and Schneider becomes equivalent to the statement that if the integral vanishes then β is rational. Again this is a statement of the shape given in the Integral Conjecture which will be stated and explained in Section 3.

C. L. Siegel mentiones in his book two problems related to what we were discussing. He first remarks that the methods developed so far do not give any non-trivial result on elliptic integrals of the third kind and secondly he mentioned the integral

$$\int_0^1 \frac{dx}{(1+x^3)} = \frac{1}{3}\left(\log 2 + \frac{\pi}{\sqrt{3}}\right).$$

He says that it is not known whether it is irrational. In 1966 in a big break-through A. Baker obtained his famous result on linear forms in logarithms from which the transcendence of the integral can be easily deduced.

The first results on periods of elliptic integrals of the third kind were obtained by M. Laurent but under strong restrictions on the the number of poles of the differential form and their residues. These restrictions were removed by again the author using totally new techniques and this gave the solution of Schneider's 3rd problem.

3. Leibniz conjecture from a modern point of view

We fix the scheme X as in 1 and denote by $\mathcal{M}_X^1$ the sheaf of local mero-morphic 1-forms on X. The mapping space $\mathcal{P}_X$ is defined to be the space of all continuous maps $\gamma : [0, 1] \to X(\mathbb{C})$ with the compact open topology. Then the integral problem can be regarded as a question on the properties of the bilinear form

$$I : \mathcal{P}_X \times \Gamma(X, \mathcal{M}_X^1) \to \mathbb{C} \cup \{\infty\} \tag{3.1}$$

given by integration $(\gamma, \xi) \mapsto \int_\gamma \xi$. Here the linearity in the first variable refers to the groupoid nature of the mapping space. The conjecture of Leibniz addresses to certain subspaces of the two factors related to the underlying problem in celestial mechanics.

Since we ask a transcendence question we have to restrict ourselves to the part $\mathcal{P}_X(\overline{\mathbb{Q}})$ of the mapping space which consists only of paths γ with $\gamma(0), \gamma(1) \in X(\overline{\mathbb{Q}})$. There is, as can be shown, a map $h_X : \mathcal{P}_X \to F_{-1}/F_0$ where $F_{-1} \supset F_0$ denotes the mixed Hodge filtration for $H_1(X, \mathbb{C})$. In the Integral Conjecture below we give a geometric description of the exceptional set $\mathcal{E} \subset \mathcal{P}_X \times \Gamma(X, \mathcal{M}_X^1)$ consisting of all pairs (γ, ξ) with $\gamma \in \mathcal{P}_X(\overline{\mathbb{Q}})$ and all $\xi \in \Gamma(X, \mathcal{M}_X^1)$ which satisfy $d\xi = 0$ and with the property that $\int_\gamma \xi \in \overline{\mathbb{Q}}$. The Integral Conjecture gives a precise geometric description of this set.

We denote by D the reduced polar divisor of ξ and by U the Zariski open set $X \setminus D$. Let $\iota : H \to H_1(U, \mathbb{Z})$ be a mixed Hodge substructure of the mixed Hodge structure $H_1(U, \mathbb{Z})$ and $\iota^\vee$ dual to ι. Then $H^\perp = \ker \iota^\vee$ is a mixed Hodge substructure of $H^1(U, \mathbb{Z}) = H_1(U, \mathbb{Z})^\vee$. The mixed Hodge structure $H^1(U, \mathbb{C})$ contains $H^0(U, \Omega_U^1)$ by Hodge theory. We define $H_{\mathbb{C}} = H \otimes_{\mathbb{Z}} \mathbb{C}$ and $H_{\mathbb{C}}^{\perp} = H^{\perp} \otimes_{\mathbb{Z}} \mathbb{C}$ and introduce the spaces $\mathcal{H}_H = \mathcal{P}_U(\overline{\mathbb{Q}}) \times_{F_{-1}/F_0} H_{\mathbb{C}}$ and $\mathcal{V}_H = H^0(U, \Omega_U^1) \times_{H^1(U,\mathbb{C})} H_{\mathbb{C}}^{\perp}$. The fiber product $\mathcal{H}_H$ is a subgoupoid of $\mathcal{P}_X$ and $\mathcal{V}_H$ is a vector subspace of $\Gamma(X, \mathcal{M}_X^1)$. The following conjecture extends the Leibniz conjecture (see also [12]).

Integral conjecture (Leibniz, Arnold). $\mathcal{E} = \cup_H \mathcal{H}_H \times \mathcal{V}_H$ *with the union taken over all the proper mixed Hodge substructures* $\iota : H \to H_1(U, \mathbb{Z})$ *as described above.*

In the integral conjecture we have to make the assumption that the differential forms are closed. However it should also hold in the general situation. Our assumption here is needed to express the differential form as a pullback of an invariant differential form on a commutative algebraic group. It would be a major progress if the condition could be removed.

We give as an illustration of the conjectural statement an example which shows how to apply the conjecture to classical questions.

Example 3.1 (Lindemann's Theorem). As an example how the integral conjecture is applied we show that $\log \alpha$ is transcendental for $\alpha \neq 0, 1$ on assuming that the integral conjecture holds. We take $X = \mathbb{P}^1$ with homogeneous coordinates x and y, we take $\xi = d \log(x/y)$ and we let γ be the path with $\gamma(0) = 1$ and $\gamma(1) = \alpha$. The divisor of ξ is $(\xi) = [0]+[\infty]$ and this shows that $U = \mathbb{A}^1 \setminus \{0\}$. The mapping space $\mathcal{P}_U$ is a fiber bundle over the universal covering space $\tilde{U}$ of U with the homotopy classes of paths as fibers. Suppose that $\log \alpha$ is algebraic. Since $H_1(U, \mathbb{Z}) = \mathbb{Z} \epsilon$ with ϵ the homology class of the path $t \in [0, 1] \mapsto e^{2\pi i t} \in \mathbb{G}_m(\mathbb{C})$ and since there are no proper and non-trivial Hodge substructures of $H_1(U, \mathbb{Z})$ the theorem implies that $\gamma = 0$. This is a contradiction.

4. The period ring of Kontsevich

Modern Hodge theory and the theory of motives or more generally of mixed motives have been a strong motivation to develop a conceptual theory of periods. A first step has been done by Kontsevich in [6] who introduced a period algebra which we briefly describe. We start with a quadruple (X, D, ξ, γ) consisting of a smooth algebraic variety X over $\mathbb{Q}$ and a divisor $D \subset X$ together with an algebraic differential form $\xi \in \Gamma(X, \Omega^d(X))$ on X of top degree which then is automatically closed and a homology class $\gamma \in H_d(X_{\mathbb{C}}, D_{\mathbb{C}}; \mathbb{Q})$. In [6], see also [7], Kontsevich defined the space $\mathcal{P}_+$ of *effective periods* as the vector space generated by the symbols $[(X, D, \xi, \gamma)]$ representing the equivalence classes

of (X, D, ξ, γ) with respect to the relation generated by the follwing "geometric relations":
- linearity in ξ and γ
- change of variables
- Stokes' formula.

Integration $\int : \mathcal{P}_+ \to \mathbb{C}$ given by $[(X, D, \xi, \gamma)] \mapsto \int_\gamma \xi$ is an evaluation map.

Kontsevich conjecture. *The evaluation map is injective.*

The conjecture implies that all algebraic relations between periods are induced by geometric relations. Unfortunately Kontsevich does not give a full construction of the space $\mathcal{P}_+$ and many details are omitted. Recently Huber-Klawitter and Müller-Stach [5] have started to modified Kontsevich's approach using and further developing Nori's theory of motives. This has quite a few advantages. In particular it is not necessary to restrict to differential forms of degree $d = \dim X$.

Example 4.1. As an illustrating example we take an elliptic curve E over $\overline{\mathbb{Q}}$. Its de Rham cohomology $H_{DR}(E)$ is a vector space of dimension 2 over $\overline{\mathbb{Q}}$ with basis ω and η. We assume that E has complex multiplication. Then the singular homology is a module of rank 1 over the endomorphism algebra End(E) of E. The subalgebra $\mathcal{P}(E)$ of $\mathcal{P}_+$ generated over $\mathbb{Q}$ by the 4 symbols $[(E, \emptyset, \xi, \gamma)]$ for $\xi = \omega, \eta$ and $\gamma = \epsilon, \delta$ is a quotient of the algebra freely generated by these symbols by the ideal generated by the graphs of the endomorphisms. This algebra is generated by two elements $[(E, \emptyset, \omega, \epsilon)]$ and $[(E, \emptyset, \eta, \epsilon)]$. If we add the symbol $[(\mathbb{P}^1, (0) + (\infty), \frac{dt}{t}, S^1)]$ as a free generator we obtain an algebra of dimension 5 and a further relation corresponding to the Legendre relation coming from Stokes Theorem. In toto the new algebra coincides with $\mathcal{P}(E)$ and a beautiful theorem of Chudnovsky says that the restriction of the evaluation map to $\mathcal{P}(E)$ is injective. This implies in particular that $\mathcal{P}(E)$ is a free algebra over $\mathbb{Q}$ of dimension 2.

5. Schanuel's conjecture

One of the most far reaching conjectures in transcendence theory is Schanuel's conjecture which, as a corollary, implies the famous theorem of Lindemann on the algebraic independence of the numbers $e^{\alpha_1}, \ldots, e^{\alpha_n}$ for any algebraic numbers $\alpha_1, \ldots, \alpha_n$ which are linearly independent over the rationals. Lindemann's theorem was the model for the conjecture which states that if $x_1, \ldots, x_n$ are complex numbers which are linearly independent over the rational numbers then the transcendence degree of the field $\mathbb{Q}(x_1, \ldots, x_n, e^{x_1}, \ldots, e^{x_n})$ is at least n. Lindemann's theorem shows that the conjecture is sharp. The Schanuel conjecture can be generalized to the exponential map of an arbitrary algebraic group G which is defined over $\overline{\mathbb{Q}}$ with Lie algebra $\mathfrak{g}$. We write $\mathfrak{g}_{\mathbb{C}}$ for the complex Lie algebra of G.

Generalized Schanuel's conjecture. *For* $u \in \mathfrak{g}_{\mathbb{C}}$, $u \neq 0$, *let* $d + 1$ *be the dimension of the smallest algebraic subgroup of* $\mathfrak{g} \times G$ *over* $\overline{\mathbb{Q}}$ *which contains the image of the 1-parameter subgroup* $\varphi : \mathbb{G}_{a,\mathbb{C}} \to \mathfrak{g}_{\mathbb{C}} \times G_{\mathbb{C}}$ *with* $\varphi(1) = (u, \exp_G(u))$. *Then the Zariski closure of* $(u, \exp_G(u))$ *over* $\overline{\mathbb{Q}}$ *in* $\mathfrak{g} \times G$ *has dimension at least d.*

It is easy to see that the conjecture implies, as an example, the elliptic analogue of Lindemann's theorem. Here the function e^z is replaced by the Weierstraß $\wp$-function $\wp(z)$ with algebraic invariants g_2 and g_3. This has been established in [8] in the case when the underlying elliptic curve has complex multiplication but is still open in the general case. Another example for an application of the Schanuel conjecture is the algebraic independence of the numbers $\log \alpha_1, \ldots, \log \alpha_n$ for $\alpha_1, \ldots, \alpha_n$ as above. This can also be deduced from Kontsevich's conjecture by considering $[(\mathbb{P}^1, (0) + (\infty), \frac{dt}{t}, S^1)]$. The result can be applied to Leopold's conjecture on the regulator and quite likely also to Beilinson regulators.

It looks as if there were a significant difference between the two conjectures. For example Lindemann's Theorem does not follow, as it seems, from Kontsevich's conjecture. It would be very interesting however to falsify this impression and a key could be the motivic Galois group which might relate the Kontsevich conjecture to the generalized Schanuel conjecture on observing that the latter covers also linear algebraic groups. Our discussion shows that Kontsevich's conjecture as well as the generalized Schanuel conjecture are extremely difficult and far out of reach in the framework of present technologies. We shall now discuss a weaker version of the conjecture for the subcategory $\mathcal{M}_1$ of 1-motives.

6. Period ring of 1-motifs

In analogy to Kontsevich's construction of the period ring we introduce in this section the period ring of 1-motifs. We take as the basic category the category $\mathcal{M}_1$ of 1-motifs à la Deligne. This is an idempotent complete additive category and its period ring $\mathcal{P}(\mathcal{M}_1)$ is constructed by taking the subcategory of $\mathcal{P}_+$ generated by $\mathcal{M}_1$ over $\mathbb{Q}$. The generators are then triples (M, γ, ξ) with M a 1-motif, $\gamma \in T(M)$ and $\xi \in T_{DR}(M)$. As relations we take

$$(M, \gamma + \gamma', \xi) = (M, \gamma, \xi) + (M, \gamma', \xi)$$
$$(M, \gamma, \xi + \xi') = (M, \gamma, \xi) + (M, \gamma, \xi') \tag{6.1}$$

If $f : (M, \gamma, \xi) \to (M', \gamma', \xi')$ is a morphism, that is f is a morphism from M to M' which satisfies $f_* \gamma = \gamma'$ and $f^* \xi' = \xi$, then we add the relation

$$(M, \gamma, \xi) = (M', \gamma', \xi'). \tag{6.2}$$

We also impose Stokes' formula, which says that

$$(M, \gamma, d\xi) = (X, \partial\gamma, \xi). \tag{6.3}$$

Here we have made the identification $X = [X \to 0]$ and this is an Artin motif which has weight 0. We should point out that the relation (6.3) referring to Stoke's theorem is in the current context a consequence of (6.2) and in principle superfluous as has been pointed out to me by J. Ayoub.

When restricted to the subring $\mathcal{P}(\mathcal{M}_1)$, the Kontsevich conjecture becomes

1-Motif conjecture. *On $\mathcal{P}(\mathcal{M}_1)$ the integration map is injective.*

The conjecture can be applied to the motif $[X \hookrightarrow E]$ with E an elliptic curve. Its periods generate a subring of $\mathcal{P}(\mathcal{M}_1)$ and if r denotes the rank of X over $\mathrm{End}_{\mathbb{Z}}(E)$ then the conjecture implies that the transcendence degree of the subring is equal to $2(r + 1)$ if E has complex multiplication and $2(r + 2)$ otherwise. From the analytic subgroup theorem (see [11], [3]) it follows that the vector space dimension over $\mathbb{Q}$ (and even $\overline{\mathbb{Q}}$)) of periods for $[X \hookrightarrow E]$ has these values (for results of this type see Section 6.2 in [3]). A calculation of the Mumford Tate group which is an extension of GL_2 by $\mathrm{Hom}(X, H_1(E))$ and, on assuming deep conjectures like the Hodge conjecture, is equal to the motivic Galois group, then shows that the linear dimension is equal to the algebraic dimension and the statement then follows from the conjecture.

The statement also follows from the generalizes Schanuel conjecture applied to an abelian variety and shows that both conjectures have a nontrivial intersection. As we have already indicated all the three conjectures are completely out of reach and of similar nature concerning depth, complexity and difficulty as the famous millennium problems. If the ground field is a function field instead of the field of algebraic numbers then Conjecture 5 has been proved by Ax [2] in the case of a torus and in the case of elliptic curves there is some work of Brownawell and Kubota [4].

If we restrict the integration map to the linear part $\mathcal{L}(\mathcal{M}_1)$ of $\mathcal{P}(\mathcal{M}_1)$ generated linearly over $\mathbb{Q}$ by the 1-motives then one can try to give an answer for a weak version of the 1-motif conjecture. At the same time it would give a weak version for the general Kontsevich conjecture since 1-motives coincide with motives attached to curves and therefore $\mathcal{L}(\mathcal{M}_1)$ is the same as the "linear part" $\mathcal{L}(\mathcal{P}_+)$ in $\mathcal{P}_+$.

Weak 1-motif conjecture. *On $\mathcal{L}(\mathcal{P}_+)$ the integration map is injective.*

If we take a toroidal motif, that is a motif of the form [X $\to$ T], then the weak 1-motive conjecture becomes Baker's Theorem on logarithmic forms. In the case of an elliptic motif [X $\to$ E] the conjecture implies [9] and if one takes an extension of an elliptic curve by a torus this covers the results of [10].

It is a great pleasure to thank here Annette Huber-Klawitter and Joseph Ayoub for extended motivic advice.

References

[1] V. I. ARNOLD, "Huygens and Barrow, Newton and Hook", Birkhäuser Verlag, Basel-Boston-Berlin, 1990.

[2] J. AX, *On Schanuel's conjectures*, Ann. of Math. **93** (1971), 252–268.

[3] A. BAKER and G. WÜSTHOLZ, *Logaritmic forms and diophantine geometry*, In: "New Mathematical Monographs" 9, Cambridge University Press, 2007.

[4] BROWNAWELL, W. DALE and K. K. KUBOTA, *The algebraic independence of Weierstrass functions and some related numbers*, Acta Arith. **33** (1977), 111–149.

[5] A. HUBER-KLAWITTER and S. MÜLLER-STACH, *On the relation between Nori motives and Kontsevich periods*, preprint (2011).

[6] M. KONTSEVICH, *Operads and motives in deformation quantization*, Letters in Mathematical Physics **48** (1999), 35–72.

[7] M. KONTSEVICH and D. ZAGIER, *Periods*, In: "Mathematics unlimited – 2001 and beyond", Springer, Berlin, 2001, 771-808.

[8] G. WÜSTHOLZ, *Über das Abelsche Analogon des Lindemannschen Satzes I*, Invent. Math. **72** (1983), 363–388.

[9] G. WÜSTHOLZ, *Transzendenzeigenschaften von Perioden elliptischer Integrale*, Journ. reine u. angew. Math. **354** (1984), 164–174.

[10] G. WÜSTHOLZ, *Zum Periodenproblem*, Invent. Math. **78** (1984), 381–391.

[11] *Algebraische Punkte auf analytischen Untergruppen algebraischer Gruppen*, Annals of Math. **129** (1989), 501–517.

[12] G. WÜSTHOLZ, *On Leibniz' conjecture, periods and motives*, in preparation.

The geometry and curvature
of shape spaces

David Mumford

The idea that the set of all smooth submanifolds of a fixed ambient finite dimensional differentiable manifold forms a manifold in its own right, albeit one of infinite dimension, goes back to Riemann. We quote his quite amazing Habilitatsionschrift:

> There are, however, manifolds in which the fixing of position requires not a finite number but either an infinite series or a continuous manifold of determinations of quantity. Such manifolds are constituted for example by the possible shapes of a figure in space, etc.

The group of diffeomorphisms of a fixed finite dimensional manifold is one such infinite dimensional manifold. The differential geometry of the subgroup of volume preserving diffeomorphisms was studied in the ground breaking paper of Arnold [1] where, in particular, he showed that its geodesics (in the simplest L^2 metric) were the solutions of the Euler equation of incompressible fluid flow. In recent years, the demands of medical imaging and, more generally, of object recognition in computer vision, have stimulated work on the space of simple closed plane curves in $\mathbb{R}^2$ and the space of compact surfaces in $\mathbb{R}^3$ homeomorphic to a sphere. One can endow these spaces with a variety of different Riemannian metrics and work out both the geodesic equation and the curvature tensor in these metrics. Many different phenomena appear giving these spaces very different characteristics in different metrics. My lecture will discuss four examples, each illustrating quite different behavior, based largely on joint work in the last ten years with my collaborators and students Peter Michor, Laurent Younes, Jayant Shah, Eitan Sharon, Matt Feiszli, Mario Micheli and Sergey Kushnarev.

1. The simplest possible example one might look at is the L^2 metric on the space of simple closed plane curves. To fix notation, let S be this space, the curves being assumed to be smooth, i.e. C^∞. Let $[C] \in S$ be the point defined by the curve $C \subset \mathbb{R}^2$. The tangent space $T_{[C]}S$ is naturally isomorphic to the space of normal vector fields to C, $\Gamma(Nor(C))$. If $\vec{n}$ is the unit outward normal and s is arc length along C, we put a metric on this via:

$$||a.\vec{n}||^2 = \int_C a(x)^2 ds(x)$$

What does S 'look like' in this metric? It is an infinite dimensional version of the string theory view of the real world: it is wrapped up more and more tightly in all its higher dimensions. In fact all its sectional curvatures are *non-negative* and go strongly to infinity in the higher frequency dimensions of the local coordinate a. However, the exponential map from the tangent space $T_{[C]}S$ to S is locally well-defined as the geodesic equation is a non-linear hyperbolic equation but conjugate points are dense on every geodesic. The global geometry collapses in the sense that the infimum of lengths of paths joining any two curves $[C_1]$, $[C_2]$ is zero. This constellation of properties seems to characterize one possible extreme in the galaxy of infinite dimensional Riemannian manifolds.

The formulas are quite simple and beautiful. The geodesic equation can be written like this. Suppose $[C_t]$ is a path in S. To describe the second derivative of the path, we can first use orthogonal trajectories to map each C_{t_0} to all nearby C_t's. Then a normal vector field $a(x, t).\vec{n}_{C_t}(x)$, $x \in C_t$ is defined by a function $a(x, t)$, $x \in C_{t_0}$ too. In particular, the tangents to the path $[C_t]$ are given near t_0 by a function of two variables $a(x, t)$, $x \in C_{t_0}, t \approx t_0$. All geodesic equations express the second derivative along a path as a quadratic function of the first derivatives. In our case, this means that the first derivative of a should be a quadratic function of a and at t_0 this is what it is:

$$\frac{\partial a}{\partial t}(x) = \tfrac{1}{2}\kappa_C(x).a(x)^2, \quad \kappa_C = \text{curvature of } C.$$

Although it may not look like it, this is a hyperbolic equation: you need only rewrite it using local equations like $y = f(x, t)$ for C_t and the curvature κ contributes an f_{xx} term with positive coefficient. This equation does seem to produce singularities in finite time: see Figure 1. Details on this and similar metrics can be found in [2–4].

The formula for curvature is even more elegant. Recall that sectional curvature is just the Riemann curvature tensor $R(a, b, a, b)$ evaluated on an orthonormal basis of a 2-plane, and that this is a quadratic form on

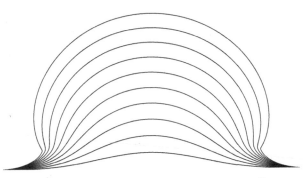

Figure 1. A geodesic in the space of plane curves in the L^2 metric. The path starts at the x-axis and moves in the direction of small 'blip'. As the blip enlarges it creates sharper and sharper corners where the curvature goes to infinity so that the geodesic cannot be prolonged.

the wedge $a \wedge b$ of its two tangent vector arguments. We will write $R(a, b, a, b)$ as $R(a \wedge b)$. So what could be more natural than:

$$R_S(a \wedge b) = \tfrac{1}{2} \int_C (ab' - ba')^2 ds \geq 0$$

where a and b define two tangent vectors in $T_{[C]}S$. The formula shows that higher frequencies produce more and more positive curvature. In fact, what happens is that path C_t in S can be shortened by adding high frequency 'wiggles' to the intermediate curves. This is illustrated in Figure 2 below.

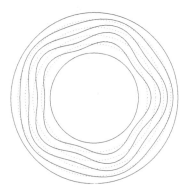

Figure 2. The set of all circles with fixed center is a geodesic in the L^2 metric if the radius varies as $t^{2/3}$. However conjugate points are dense on it: Here is a deformation of this geodesic which has a conjugate point when the radius increases by the factor 1.8957.... Beyond that point, it shortens the length of the geodesic.

To summarize: the geodesic equation is a non-linear hyperbolic PDE with well posed initial value problem; the curvature is non-negative, going strongly to infinity at high frequency and with conjugate points dense; and the global metric is identically zero because the infimum of path lengths is zero. This behavior is typical of L^2 metrics.

2. Positive curvature which is, however, tamer is produced in another elegant situation. This example is due to the work of Laurent Younes [5, 6]. Here we regard the plane as the complex plane. The remarkable idea is to consider the complex square root of the derivative of the curve, i.e. if $t \mapsto f(t) \in \mathbb{C}, t \in \mathbb{R}/2\pi\mathbb{R}$ is the curve, define $g(x) + ih(x) = \sqrt{f'(t)}$. If C is an embedded curve (or more generally any immersed plane curve with odd index), then $g(x + 2\pi) \equiv -g(x), h(x + 2\pi) \equiv -h(x)$. The closedness of the curve is expressed by the formula $\int_0^{2\pi} f'(t)dt = 0$ which means that in $L^2([0, 2\pi])$ g and h are orthogonal functions of the same length. We can reverse this process and, starting from such a pair g, h, define a parameterized curve, up to translation, by:

$$x \longmapsto f(x) = \int_0^x (g(x) + i.h(x))^2 dx.$$

The upshot of this ansatz is this: Let $\mathcal{H}$ be the Hilbert space of functions g such that $g(x + 2\pi) \equiv -g(x)$ with norm $||g||^2 = \int_0^{2\pi} g(x)^2 dx$. Let $\mathcal{G}$ be the Grassmannian of 2-planes in $\mathcal{H}$ and let $\mathcal{G}_0$ be the open subset of 2-planes such that there is no x where all functions in the 2-plane vanish. Then using orthonormal bases of these 2-planes as g and h, we find that $\mathcal{G}_0$ is isomorphic to the space of parameterized immersed plane curves of odd index mod translations, rotations and scaling. Not only that but the natural metric on this Grassmannian corresponds to a very natural metric on this space of curves. The tangent space to parameterized curves is given by all vector fields along the curve, not merely those which are normal, thus, in our case, by a complex valued function along C. The Grassmannian metric turns out to equal the 1-Sobolev norm (with only first derivatives):

$$||a||^2 = \frac{1}{\text{len}(C)} \int_C |a'(x)|^2 ds(x), \quad s = \text{arc length}.$$

Geodesics and curvature on a Grassmannian are given by quite simple and classical formulas so we also get formulas for these both on this space of *parameterized* curves and on its submersive quotient of *unparameterized* immersed curves both mod translations, rotations and scalings. The geodesic equation is now an integro-differential equation most

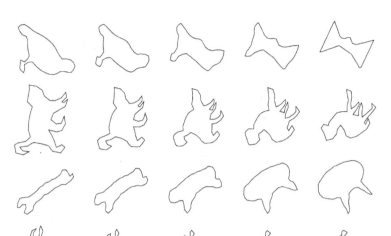

Figure 3. Some geodesics in Younes's metric between plane curves representing recognizable shapes. Note how they rotate to make optimal matches, e.g. the tail of the cat with the head of the camel.

easily written not in terms of velocity a but in terms of a 'momentum' which is the second derivative $u = -d^2a/ds^2$. Like the Grassmannian itself, these spaces also have entirely non-negative curvature but not so strongly positive that this prevents the Riemannian metric from defining a nice global metric. The space has finite diameter in its global metric and can be completed by adding certain non-immersed curves. Some examples of geodesics in this space are shown in Figures 3 and 4. This type of space seems to be the natural infinite dimensional analog of compact symmetric spaces of finite dimension.

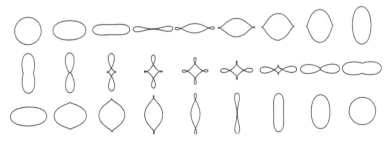

Figure 4. If we allow paths to pass through some non-immersed curves, we find many closed geodesics in this metric. This is the simplest example. The path goes from left to right, row by row; a loop flips over in the ellipse-like shapes at the end of the first row and the beginning of the third, hence these are non-immersed.

3. A third metric can be put on simple closed plane curves, here modulo translations and scalings, but now with *non-positive* curvature. Interestingly, only one half a derivative is added to the metric in the previous example: it is Sobolev with 3/2 derivatives. This is the famous Weil-Peterson metric. It is defined as follows: start with the space of vector fields $v(\theta)$ on the circle and put the WP-norm on it, defined in terms of its Fourier transform by:

$$||v||^2_{WP} = \sum_{n=2}^{\infty}(n^3 - n)|\hat{v}_n|^2.$$

Now SL_2 is a subgroup of the group of diffeomorphisms of the circle with lie algebra consisting of the vector fields $(a+b\cos(\theta)+c\sin(\theta))\frac{\partial}{\partial\theta}$. This is clearly the null space of the above WP norm and since – miraculously – the WP-norm is also invariant under the adjoint action of SL_2, this norm extends by right translations to an invariant Riemannian metric on the coset space $SL_2\backslash\text{Diff}(S^1)$. Now the final link: this coset space is isomorphic to the space of simple closed plane curves mod translations and scalings. This comes via 'welding': given a diffeomorphism φ, attach two unit disks to each other along their boundaries using the twist φ. The result is a simply connected compact Riemann surface, hence it must be conformal to the sphere. The image of the welded common boundary is our curve. For details, see [7, 9].

One of the remarkable consequences of this construction is that it defines an operation of composition between plane curves. The welding operation also defines a bijection between the group $\text{Diff}(S^1)$ itself and triples $(C, P, \vec{t})$ where C is a simple closed plane curve, P a base point inside C and $\vec{t}$ is a distinguished ray at the base point, all modulo translations and scalings. Thus there is a law of composition of such triples. Some examples are shown in Figure 5.

This metric is the closest to the standard metric on $\mathbb{R}^n$ because (a) it is invariant under the transitive action of a group, here $\text{Diff}(S^1)$ and (b) it is quite flat in high frequency dimensions because the Ricci curvatures (which are the sum of sectional curvatures $R(a \wedge b_i)$ where $\{b_i\}$ are an orthonormal basis of $a^{\perp}$ for variable a) are known to be finite. It is also a complete complex Kähler-Hilbert manifold and has unique geodesics between any two points [7, 8]. The metric can also be defined using potential theory which embeds the curve in field lines and thus endows its interior and exterior with a rich additional structure. The geodesic equation is an integro-differential variant of Burger's equation involving the (periodic) Hilbert transform. Among geodesics on this space, there is a special class of soliton-like geodesics, which Daryll Holm named 'tei-

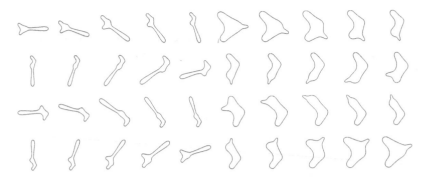

Figure 5. A selection of plane curves obtained by composing two diffeomorphisms corresponding to (i) a boomerang-like shape with base point in the middle and (ii) a finger-like shape with base point near one end respectively. In each panel of 15 curves, the data $\vec{t}$ is varied or, equivalently, a variable rotation is added in the middle of the composition.

chons'. They are the geodesics generated by vector fields v dual in the WP norm to sums of delta functions, i.e.

$$\langle v, u \rangle_{WP} = \sum_i p_i u(\theta_i), \quad \text{for all } u$$

for some p_i, θ_i. An example of a teichon is shown in Figure 6.

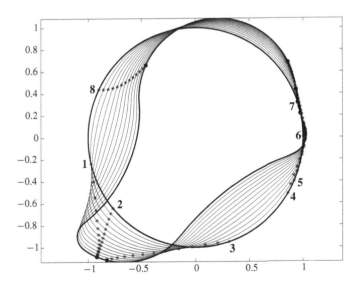

Figure 6. A geodesic from the unit circle to a duck like shape using an 8-Teichon. The figure is due to S. Kushnarev [10].

4. The final example is much more general and deals with the full diffeo-morphism group of $\mathbb{R}^n$. Arnold's curvature formula for volume preserving diffeomorphisms was significantly more complicated than anything in the above examples. In his case, there are both positively and negatively curved sections and this also seems to happen for Riemannian manifolds constructed from any higher order Sobolev type metrics on diffeomorphism groups. I would not be surprised if at some point understanding these more complex curvature formulas gives new insight into the unsolved problems of fluid flows.

The situation that my group has studied most intensively is the metric induced on 'landmark space', that is simply the space $\mathcal{L}_{n,N}$ of distinct N-tuples of points in $\mathbb{R}^n$. Fixing a base N-tuple, we get a submersive map from $\text{Diff}(\mathbb{R}^n)$ to $\mathcal{L}_{n,N}$. We may put a Sobolev norm on vector fields X,

$$\|X\|^2 = \int_{\mathbb{R}^n} \langle X, LX \rangle dx_1 \cdots dx_n$$

where L is a positive definite self-adjoint operator, e.g. $L = (I - \Delta)^s$. This defines a metric on the group of diffeomorphisms provided that L has enough derivatives. In fact, we want the finiteness of the metric to force the diffeomorphisms to be C^1. Then we get an induced Riemannian structure on the quotient space $\mathcal{L}_{n,N}$. It has a simple form. If G is the Green's function associated to L, $\{P^1, \cdots P^N\} \in \mathcal{L}_{n,N}$ and v^a is a vector at P^a, then the metric is:

$$\|\{v^1, \cdots, v^N\}\|^2 = \sum_{1 \le a, b \le N} (G^{-1})_{ab} \langle v^a, v^b \rangle, \quad G_{ab} = G(P^a - P^b).$$

Arguably this is the most natural class of metrics to put on landmark space.

The geodesic equation on landmark space is quite elegant. To any geodesic there is a natural set of momenta u_a for which a geodesic is a solution of:

$$\frac{du_a}{dt} = -\sum_b \nabla G(P^a - P^b) \langle u_a, u_b \rangle$$

$$\frac{dP^a}{dt} = \sum_b G(P^a - P^b) u_b.$$

These equations create a world governed by a weird sort of physics in which points moving together attract and points moving in opposite directions repel and occasionally one even gets planetary systems. Since $\mathcal{L}_{n,N}$ is a submersive quotient of the diffeomorphism group, geodesics in $\mathcal{L}$ lift to horizontal geodesics in the group. So these geodesics induce warpings of the ambient Euclidean space. In fact any geodesic in the diffeomorphism group can be approximated by one of these landmark geodesics if we take enough landmark points. Some examples are shown in Figure 6.

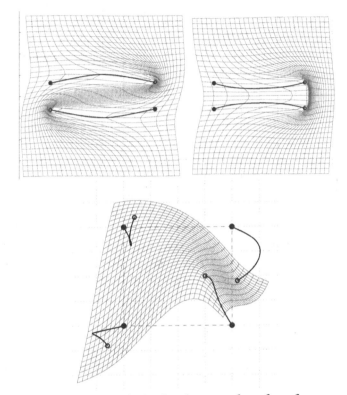

Figure 7. Three geodesics on the landmark spaces $\mathcal{L}_{2,2}$, $\mathcal{L}_{2,2}$, $\mathcal{L}_{2,4}$ respectively, plus the induced diffeomorphism of the ambient plane. Note how points moving in the same (resp. opposite) direction are drawn together (resp. pushed apart). In the four point case, this causes some complex gyrations. These figures are from the thesis of M. Micheli [11]

The sectional curvature of landmark space has four terms, similarly to Arnold's formula for the curvature of Lie groups [1]. The reason for this seems to be that it reflects several different modes of interactions of the points. There are, in particular, at least two 'causes' of positive curvature. One is seen in the middle panel of Figure 4: when two points must move in a similar direction, it saves energy for them to come close to each other. But if the distances are in a certain range, there will be two geodesics joining the pair at the initial position on the left and the same pair translated to the final position on the right. One geodesic has the points moving nearly independently and nearly parallel and the other has them first coming close, then moving together and finally moving apart again. This non-uniqueness causes positive curvature.

Another cause of positive curvature occurs in dimension three and higher. Suppose two points are to be interchanged, the first moving to the position of the second and the second to that of the first. Since the distance is infinite if they were to move directly towards each other, they must move around each other and there are many planes in which to do this. We show in [12] that when only two points have momenta, these are, in a sense, the only ways positive curvature can arise.

But negative curvature arises all the time from the turbulence caused by landmark point motion. Take the situation where a single point P is moving with non-zero momentum but that there are many other landmark points around it with zero momentum. These extra points are dragged along, compressed together in front of P. If P moves from A to B, we wind up with a configuration $C(B)$ of the whole set of landmarks. Take $B_1 \neq B_2$. Then what will the geodesic from $C(B_1)$ to $C(B_2)$ look like? You can't just put momentum on P because you need to move the points bunched up near B_1 back apart and create a new bunch near B_2. The only way to do this is unwind the mess you made in one geodesic and recreate the new mess in the second. This is negative curvature: to connect the endpoints of two trips, it is better to go nearly back home. For details, see [12].

Anyway, the somewhat daunting formula for sectional curvature is this. Let $v_1 = \{v_1^a\}$ and $v_2 = \{v_2^a\}$ be two tangent vectors to $T\mathcal{L}$ at some point $\{P^a\}$. Let v_i be extended to a vector field on $\mathbb{R}^n$ by its horizontal lift to the diffeomorphism group. Let v_1^b and v_2^b be the co-vectors dual to the v's. Then the numerator of sectional curvature is given by:

$$R(v_1 \wedge v_2) = R_1 + R_2 + R_3 + R_4$$

$$R_1 = \frac{1}{2} \sum_{a \neq b} \left[(v_2^b)_a \otimes \delta^{ab} v_1 - (v_1^b)_a \otimes \delta^{ab} v_2 \right] \cdot H^{ab}$$

$$\cdot \left[(v_2^b)_b \otimes \delta^{ab} v_1 - (v_1^b)_b \otimes \delta^{ab} v_2 \right],$$

$$R_2 = \langle D_{11}, F_{22} \rangle - \langle D_{12} + D_{21}, F_{12} \rangle + \langle D_{22}, F_{11} \rangle,$$

$$R_3 = \| F_{12} \|_{T^*\mathcal{L}}^2 - \langle F_{11}, F_{22} \rangle_{T^*\mathcal{L}},$$

$$R_4 = -\frac{3}{4} \| D_{12} - D_{21} \|_{T\mathcal{L}}^2,$$

where $\delta^{ab} v = v^a - v^b$, and $C_{ab}(v) = \langle \delta^{ab} v, \nabla G(P_a - P_b) \rangle$ for any v and $D_{ij}^a = \sum_{b \neq a} C_{ab}(v_i)(v_j^b)_b \in T\mathcal{L}$,

and $(F_{ij})_a = \frac{1}{2} \left(D_a v_i \cdot (v_j^b)_a + D_a v_j \cdot (v_i^b)_a \right) \in T^*\mathcal{L}$, ($D_a =$ deriv. at P^a) and $H^{ab} = I \otimes D^2 G(P^a - P^b)$.

The term R_4 above is the main cause of the turbulence related negative curvature: it is the only term which involves points with no momentum of their own. It is natural to generalize this formula to get more insight into it. A paper is under preparation analyzing the spaces of submanifolds of any type in any fixed ambient finite dimensional manifold M with respect to a very general Sobolev-type metric on the group of diffeomorphisms of M.

References

[1] V. ARNOLD, *Sur la géométrie différentielle des groupes de Lie de dimension infinie et ses applications l'hydrodynamique des fluides parfaits*, Annales de l'institut Fourier **16** (1966), 319–36.

[2] P. MICHOR and D. MUMFORD, *Riemannian Geometries on Spaces of Plane Curves*, J. of the Europ Math. Society **8** (2006), 1–48.

[3] P. MICHOR and D. MUMFORD, *Vanishing geodesic distance on spaces of submanifolds and diffeomorphisms*, Documenta Mathematica, 10, 2005.

[4] P. MICHOR and D. MUMFORD, *An overview of the Riemannian metrics on spaces of curves using the Hamiltonian Approach*, Applied and Computational Harmonic Analysis **23** (2007), 74–113.

[5] L. YOUNES, *Computable elastic distances between shapes*, SIAM J. Appl. Math. **58** (1998), 565-586.

[6] P. MICHOR, D. MUMFORD, J. SHAH and L. YOUNES, *A metric on shape space with explicit geodesics*, Rendiconti Lincei – Matematica e Applicazioni **19** (2009), 25–57.

[7] T. TAKHTAJAN and L.-P. TEO, "Weil-Petersson Metric on Universal Teicmüller Space", Memoirs of the AMS, Vol. 86, 2008.

[8] F. GAY-BALMAZ, "Infinite Dimensional Flows and the Universal Teichmüller Space", PhD thesis, École Polytechnique de Lausanne, 2009.

[9] D. MUMFORD and E. SHARON, *2D-Shape Analysis using Conformal Mapping (with Eitan Sharon)*, Int. J. of Comp. Vision **70** (2006),55–75.

[10] S. KUSHNAREV, *Teichon: Soliton-like Geodesics on Universal Teichmüller Space*, J. Exp. Math. **18** (2009), 325-336.

[11] M. MICHELI, "The Differential Geometry of Landmark Shape Manifolds", PhD thesis, Div. of Appl. Math., Brown Univ., 2008.

[12] M. MICHELI, P. MICHOR and D. MUMFORD, *Sectional curvature in terms of the Cometric, with applications to the Riemannian manifolds of Landmarks*, submitted.

COLLOQUIA

The volumes of this series reflect lectures held at the "Colloquio De Giorgi" which regularly takes place at the Scuola Normale Superiore in Pisa. The Colloquia address a general mathematical audience, particularly attracting advanced undergraduate and graduate students.

Published volumes

1. Colloquium De Giorgi 2006. ISBN 978-88-7642-212-6
2. Colloquium De Giorgi 2007 and 2008. ISBN 978-88-7642-344-4
3. Colloquium De Giorgi 2009. ISBN 978-88-7642-388-8, e-ISBN 978-88-7642-387-1